ESSAYS

ON THE

SPIRIT OF THE INDUCTIVE PHILOSOPHY

ESSAYS

ON THE

SPIRIT OF THE INDUCTIVE PHILOSOPHY,

THE

UNITY OF WORLDS,

AND THE

PHILOSOPHY OF CREATION.

BY

THE REV. BADEN POWELL,

M.A. F.R.S. F.R.A.S. F.G.S.

SAVILIAN PROFESSOR OF GEOMETRY IN THE UNIVERSITY OF OXFORD.

LONDON:

LONGMAN, BROWN, GREEN, AND LONGMANS.

1855.

S.B.N. - GB: 576.29160.9

Republished in 1969 by Gregg International Publishers Limited
Westmead, Farnborough, Hants., England

Printed in offset by Franz Wolf, Heppenheim/Bergstrasse
Western Germany

PREFACE.

THE three following essays, though somewhat connected in subject, are yet each distinct and complete in themselves, having been originally composed at different times and with separate objects. Hence there will probably be found in some parts *repetitions:* but on the whole it appeared preferable to allow these to remain, rather than by omissions and alterations to render less complete and continuous the argument of each essay in itself. And the few topics which belong to them in common will, in most cases, be found treated under somewhat different aspects, according as the particular argument in each instance required.

The First Essay consists mainly of an amplification of a few paragraphs in my paper "on Necessary and Contingent Truth" in the Oxford Ashmolean Memoirs, 1849, in reference to which I felt it desirable to explain and illustrate some points there but imperfectly treated, as well as some other topics related to them, and which have of late years been the subject of considerable discussion; some of which were also considered in my work on "the Connexion of Natural and Divine Truth," 1838. More precisely, the subjects of the primary grounds of inductive reasoning, and the theory of Causation, have long since appeared to me to be commonly involved in much confusion of thought, which has, as I think, been rather increased than diminished by some recent discussions from which we might have hoped for greater enlightenment; — and which appears to me to be the source of many unhappy difficulties and objections connected with the so-called doctrine of "final causes," and the evidences of natural theology generally.

To the object of clearing up some of these difficulties, and inculcating better views, some parts of my former work last referred to were devoted: And to the argument there pursued (so far as I am aware) no substantial objections have been alleged. Yet the frequent reproduction of the same original confusion of language and thought, in otherwise able and valuable writings at the present day, renders it not useless to recall attention to some of those considerations by which, I believe, the whole subject is put on a more satisfactory and unobjectionable basis.

Many of these topics, it will be evident at first sight, are coextensive with those so elaborately and profoundly treated in Dr. Whewell's Philosophy of the Inductive Sciences and in Mr. Mill's Logic. If I have made very few specific references to either of those treatises, it has arisen from no want of respect and consideration for either of the distinguished authors; but rather from an opposite feeling of high general esteem for the ability with which they have treated the subject, I entertained an unwillingness to appear to enter into direct controversy, in some material questions on which I have been constrained to hold opinions somewhat differing from those of both writers.

If the grounds on which I maintain my views shall be found sufficiently indicated and explained, I trust the candid reader will be as well prepared to come to an unbiassed opinion on the points in question as if they were urged with a greater degree of critical detail; and the opinions which I controvert will be equally marked out, without more minute reference to the particular authors.

The Second Essay was called forth by a perusal of the two able and interesting works on the question of the Plurality of Worlds, which have of late attracted such an unexpected degree of public attention; an interest which, even up to the moment of bringing out this volume, does not appear to have abated, if we may judge from the several other publications since announced on the same question.

With respect to the author of the "Essay on the Plurality of Worlds," while it would be absurd to pretend ignorance of his real eminence, I have throughout felt it would at the same time be improper to refer to his opinions, otherwise than as those sustained by the masked character under which, doubtless for the greater freedom of such discussion, he has thought fit to veil academical dignity.

The controversy itself, as to the question of inhabited worlds, appears to me of comparatively little moment: it is rather for the sake of more general considerations involved, that I have been led to enter into the discussion, and, in some measure, to hold the balance between the two disputants. Those broader principles are closely connected with the subject of the First Essay.

The collateral questions introduced into the Second Essay have also an immediate bearing on the subject of the Third. The inquiry into the *present* condition of planetary worlds is closely connected with that of their *past* state and probable origin; and this with the general question of the history of *creation,* so far as it can be traced on physical grounds. But this subject, again, is one which has of late years extensively occupied the public attention; especially from the extraordinary popularity attained by the "Vestiges of the Natural History of Creation," and the controversies to which that work has given rise. In those controversial discussions, it cannot but be matter of regret that so acrimonious a tone, little suited to eliciting the truth, should have been adopted by some of the writers. Hence it seemed to me that a more calm and philosophical analysis of the whole question was much needed; and in some measure to supply such a review of the general principles and grounds on which all speculations of the kind should be conducted, as well as to examine dispassionately into the alleged religious bearings of any theories by which some part of the steps and processes of creation might be explained, has been the aim of the Third Essay.

It should perhaps be expressly observed that if, in those passages where I have spoken of the evidences of natural theology, I have professedly restricted my remarks to the *physical* portion of the argument,—it is not from at all disparaging or overlooking the *moral* and *metaphysical* portions, that I have not adverted to them, but solely because they are not immediately connected with the more direct object of these Essays, already extended, perhaps, to too great a length.

A similar remark ought, also, to be made with respect to the very brief and inadequate mention made of some other points of deeper import to the belief in revelation; to which I could willingly have devoted a more extended discussion than it was possible within my present limits to give them.

ANALYTICAL TABLE

OF

CONTENTS.

ESSAY I.—ON THE SPIRIT OF THE INDUCTIVE PHILOSOPHY.

§ I.—The Inductive Principle.

§ II.—The Unity of Sciences.

ESSAY II.—ON THE UNITY OR PLURALITY OF WORLDS.

§ I.—The Argument considered in a Physical and Philosophical Point of View.

§ II.—The Argument considered in a Theological Point of View.

ESSAY III.—ON THE PHILOSOPHY OF CREATION.

§. I.—THE EVIDENCE DERIVED FROM GEOLOGY.

§ II. — The Evidence derived from Physiology.

§ III. — General Considerations arising out of the preceding Evidence.

§ IV.—THE BEARING OF THE PRECEDING ARGUMENTS ON THE THEOLOGICAL VIEW OF CREATION.

APPENDIX.

ERRATA.

Page 70. line 2. from bottom, for "and end" read "an end."
" 77. " 1. for "concerned" read "properly such."
" 123. (margin), for "no" read "a."
" 240. line 3. for "bind" read "bend."
" 295. " 2. for "threatened to" read "impending on."

ESSAY I.

ON THE SPIRIT OF

THE INDUCTIVE PHILOSOPHY.

§ I.—THE INDUCTIVE PRINCIPLE.

GALILEO, 1590.

"Opinionum commenta delet dies,
Naturæ judicia confirmat."—CIC.

Introductory remarks.

THE characteristic nature, genius, and grounds of the inductive philosophy have been much discussed of late years, and under considerable varieties of view, by different parties. Whilst some have carried out their view of its principles into metaphysical abstractions often hardly intelligible, others have sought to narrow them to the results of mere sensible experience; and whilst the one would connect its aims with a higher intellectual philosophy, even verging on the mystical, the other school would lower its objects to the mere empire over matter, and the attainment of utilitarian ends.

Distinction of sensations and ideas.

More precisely, an inquiry into the essential grounds and principles of induction involves the more general question of what has been termed "the fundamental antithesis"* of sensations and ideas, facts and theories; in a word, of two essentially distinct and independent sources of all knowledge, the external and the internal — observation by the senses, and ideas originating in the mind itself; while it is only by the application of the latter to reduce to system the materials supplied by the former, that any real philosophical theory can be constructed, the crude results of observation be converted into an inductive theory, or sense elaborated into science.

Thus ideal conceptions, the pure offspring of mind, the mere creatures of intellect, seem to exercise a sort of plastic power over the mass of

* See Dr. Whewell's two able memoirs "On the Fundamental Antithesis of Philosophy." Cambridge Phil. Society Transactions, 1848.

material results, giving them a fresh character and scientific significance; and thus we are enabled to make that ascent from facts to laws, from laws to causes, which is the aim and boast of the inductive philosophy. Such views, carried out in some instances to speculations of a kind still more remote and hardly comprehensible, have been adopted by many at the present day: while, on the other hand, the "positive philosophy" is characterised by a tendency to the contrary extreme of discarding all reference to those higher intellectual principles, reducing all science to the naked results of observation and calculation, and all idea of causation to that of mere invariable sequence of phenomena.

Meaning of the term "experience."

In looking more precisely to the meaning of the term *experience*, if we understand it literally as the mere collection of facts, such as sense and observation *directly* furnish, and the rejection of everything which is not, in this restricted sense, properly learnt by it, then, indeed, there is an end put to all really scientific or philosophical investigation; and beyond the narrow circle of those facts we can never enlarge our conceptions or raise our contemplations.

The slightest consideration, however, will show

Induction implies more than sensible experience.

that the term *experience*, even in the simplest case, must be understood in a wider sense: whilst the logical analysis* of induction exhibits a syllogism in which a large assumption is necessarily implied, beyond and independent of any accumulation of facts. Thus every induction is seen essentially to involve a *certain amount of hypothesis*,—a certain assumption of more than the bare facts themselves seem strictly to warrant. We form intellectual conceptions of a nature more general than the mere enumeration of a number of instances, however many; and thus supply "the string on which" (as Dr. Whewell happily expresses it) "the pearls are hung;" and perceive, according to the illustration of another able writer†, how "philosophy proceeds upon a system of credit, and that, if she never advanced beyond her tangible capital, her wealth would not be so enormous as it is."

It is certainly not the mere *number* of instances which constitutes the strength of an inductive conclusion; but it is the *kind* and *quality* of them, as

* See Archbishop Whately's Logic, book iv. ch. i. §§ 1. 2.
† Outlines of the Laws of Thought (p. 312.), by Rev. W. Thomson, M.A.: London, 1849.

bearing on the manifestation of the existence of certain relations among them, connecting them together by *analogy*. If the individual facts be thus connected, or of the *right sort*, a comparatively *small number* of them will be convincing, when in other cases the most laborious accumulation will be fruitless and unsatisfactory, as wanting in a real connection of analogy. When, however, that essential condition is secured, it then infallibly happens (as has been well said) that a "vague and local idea . . . . passes through the mint of a very few decisive experiments into the treasury of accepted truths."*

What it is which is superadded to sensible facts.

In arriving at any general inductive conclusion, then, *something* is clearly superadded to the mere mass of facts; the question is, *what* is it? In the simplest case, that of knowledge acquired by the senses, something more than mere sensation is implied: besides sensations conveyed *to* the mind, there must be corresponding ideas excited or formed *in* it. All observation which involves *mind* involves *theory:* the *facts* of sense must be *idealised.* Of

* Rev. W. V. Harcourt's Letter, &c., Phil. Mag. 1846, p. 76.

the truth of this no reflecting person entertains a doubt: the sole question is, *how* it is effected, and *whence* these ideas are formed.

Supposed inherent faculty or principle.

According to one school, these phenomena are referred to a peculiar principle, supposed to be implanted in the mind, not to be further analysed: a special faculty, producing a distinct mode of conviction; a kind of assurance, prior to and independent of external sense, and derived from the interior resources of reason; an inherent intellectual element, which warrants us in extending our conclusions beyond the mere limits of observation, and in inferring intuitively and certainly the future or unknown from the past or known. Or, more precisely, certain fundamental conceptions are supposed primarily and originally formed within the mind itself, derived somehow from its interior resources, without any reference to external sensation; and the introduction of these conceptions (differently modified according to the nature of the respective subjects) impresses the proper form on the collected facts. And it is from the fundamental ideas thus entering into combination that the attributes of universality and necessity are acquired by our conclusions and belief, and a certainty attained on *à priori* grounds which no mere observation could furnish.

Another view. Gradual process of abstraction and generalisation.

Another school, discarding all reference to any intuitive or internally created ideas, analyses the intellectual process into its elements, and shows that through successive steps of abstraction, from the simple collection of facts, we advance to theories, which are true just in proportion as we are guided by the right perception of analogy and the important rule of correcting one generalisation by another, and thus, that all knowledge is ultimately derived from observation.

Idea of intuitive principles natural.

The theory of intuitive or internal principles undoubtedly appeals powerfully to the imagination. Nothing seems more natural or plausible than to refer everything to *ultimate principles* originating in the mind: it saves the labour of further analysis, and supplies a specious explanation of intellectual phenomena, which seems to gratify at once the desire of penetrating the secrets of our nature and the love of the mysterious, in appealing to great but hidden causes within us: a species of occult philosophy, which seems eminently to harmonise with the mysticising tendencies of the age; but which, nevertheless, appears to be conceived in a spirit very opposite to that of the simple and *positive* character of the inductive method, and, though sanctioned by great names, seems rather to be a retrograde movement, and to evince a lingering attachment to the scholastic mysticism, or to be in some sense a revival of it.

Ought to be analysed up to simpler elements.

That we are *naturally prone* to entertain such notions may be very true; yet it may happen in this, as in many other instances of what we are *prone* to do, that we do *wrong*. But the more strict metaphysical inquirer will acknowledge that it is unphilosophical to imagine peculiar and unknown mental principles, if processes carried on through already acknowledged intellectual powers can be shown to suffice for explaining the facts.

In the present instance, indeed, as in other inquiries, it may be perfectly allowable in the first instance to set down any outstanding class of phenomena as *provisionally* something *sui generis*, and of an elementary character, just as in chemistry we may regard any new substance as *elementary* while it is as yet *undecomposed;* but still it is the *aim* of the chemist to decompose it if he can. In the same way there may be a multitude of ideas, impressions, intellectual sensations, and the like, which may at first seem like elementary principles; but which, nevertheless, it should be the aim of the metaphysical analyst to reduce into their component simpler elements if possible.

Power of imagination.

In such cases the powers of imagination may be appealed to; and doubtless those powers are sufficiently prolific in suggesting theories. The minds of the ancient philosophers teemed with speculative schemes of nature before any study of facts had furnished them with substantial materials. Humboldt has well observed, that "long before the discovery of the New World it was thought land could be seen in the west from the Canaries and the Azores. They were phantasms not produced by any extraordinary refraction of the rays of light, but merely by a longing for the distant, for that which lies beyond the present. The natural philosophy of the Greeks, and the physics of the middle ages and even of much later centuries, presented swarms of such fantastic forms to the imagination. The mental eye still essays to pass the horizon of limited knowledge even as the material eye endeavours to pierce the natural horizon from an island height or shore. Faith in the unusual and wonderful gives definite outlines to every product of imagination; and the realm of fancy, a strange land of cosmological, geognostial, and imaginative dreams, is incessantly blended with the world of reality."*

Yet mere imagination, however powerful and prolific, will avail little for creating any theories which will stand the test of observation, or which have any real application in nature.

Something more than imagination.

But, from considering the nature of our generalisations, it is argued, that we must necessarily *obtain ideas* from *some other source* than *sense*, or that the mind possesses a peculiar power or faculty of acquiring a higher degree of certainty from *within* than experience can give from *without*. Or, again, it is said, in such cases as mathematical theorems, the mind attains certainty quite independently of experience; whilst in other cases, such as limited inductions in subjects little known, it has no certainty beyond the mere facts which are directly presented to it. Why, then, is the mind

* Cosmos, p. 84., 1st trans.

so confident in one case and so cautious in the other, unless there be a real difference in the faculties brought into play in the two respective cases?

Logical analysis of induction.

When we analyse the process logically, it is manifest that, in induction, what is superadded to a mere collection of facts consists precisely in the assumption "*that all phenomena of the kind in question are similar to the few actually examined.*"

Origin of this assumption.

The question, then, is reduced to this, How does the mind come to make this universal assumption, and to be so firmly convinced of its truth?

Mental processes carried on unconsciously.

In the first place, I think it will be allowed, on reflection, that general conceptions of this kind, however apparently abstract in their nature, may be created in our minds by very simple causes, of whose operation we may yet be quite unconscious. There is nothing of which we are less conscious than the acquisition of the commonest ideas by daily experience, and the successive and gradual generalisation of that experience by the process of *abstraction;* and in this way we constantly obtain (without being aware of it) numberless prepossessions and convictions far stronger than any systematic demonstrations can supply.

Assumption of the uniformity of nature.

The primary assumption involved in all induction is the *presumed* uniformity of phenomena, or the conformity of other facts of the same class with that under examination to the same law or type.

Idea of generalisation derived from gradual experience.

It is, then, perfectly true that no inductive process can advance without *the assumption of this generalising principle, which is, nevertheless, antecedent to the particular class of experimental testimonies* IN THAT INSTANCE *appealed to.* But what I would particularly dwell upon is, that *it is not antecedent to* ALL *experience;* it is some principle already established in the mind by previous abstractions, *remotely* derived from previous experience, and *specially extended by* ANALOGY beyond the precise limits of actual observation in this instance.

Proneness to hasty generalisation.

It is true that there exists in the human mind a strong natural propensity to draw *hasty inferences,* to generalise too rapidly, and to deceive ourselves by erecting conclusions on very unsubstantial and insufficient data; and this is closely associated with the fondness for tracing resemblances; being pleased with uniformity and the contemplation of analogy, real or imagined, where there are often but slight indications of it, or even where the appearances of it are in reality altogether fallacious.

These propensities are evinced more or less strongly in different minds in the earliest exercise of their powers: and though in matters of common life and every-day occurrence they are soon and effectually subjected to the corrective process of enlarging experience and reflection, which the pressing necessities of daily existence force upon us; yet in other subjects, such as those of abstract speculation or philosophical inquiry, it may be long before they receive so salutary a check, or at least before they come to be really well regulated by rational principles.

Generalisations imperfect at first, corrected by increasing experience.

Our FIRST *inductions are* ALWAYS IMPERFECT AND INCONCLUSIVE; we advance towards real evidence by successive approximations; and accordingly we find false generalisation the besetting error of most first attempts at scientific research. The faculty to generalise accurately and philosophically requires large caution and long training; and is not fully attained, especially in reference to more general views, even by some who may properly claim the title of very accurate scientific observers in a more limited field. It is an intellectual habit which acquires immense and accumulating force from the contemplation of wider analogies; and in any one case our conviction of inductive truth is largely built up on past trial of its soundness in other cases; and from the perpetual multiplication of such cases it obtains a perpetually progressive character of greater certainty, increasing in a rapidly accelerated ratio as experience enlarges.

By trial of theoretical suggestions in succession, and only after repeated failure, we learn their erroneous nature. But thus by acquiring more caution and confidence and adopting better conjectures, we revise and amend our attempts, and learn to proceed on more sound principles, until we gain a habit of generalisation worthy the name of inductive power.

Tendency evinced in different degrees.

Again, the tendency to make the primary inductive assumption, and the extent to which it reaches, admit of many degrees. It is found in its higher perfection in those comprehensive views which constitute the discoveries of the greatest philosophers, and in varied inferior degrees in other instances.

Feeble in earlier stages.

In the order of time, also, it is always evinced with far less effect in the earlier stages of scientific development, and with more full and perfect force in its later progress; whether in the infancy of science, or of the experience of an individual.

Gains strength by advance of experience.

But as the cultivation of inquiry advances, the inductive process by habitual exercise derives force so naturally and insensibly, that the mind is utterly unconscious of its acquirement; and hence it is that we readily give way to the very natural, but mistaken persuasion, that the generalised idea is something inherent, or created out of the intrinsic powers of reason itself.

And from absence of contradictory cases.

And in any case even of the most limited induction, there is one argument on which, more than any other, we always fall back with perfect confidence, and which really constitutes the main force of the evidence, viz. the assurance that *if there be any fatal exception* to the law or truth supposed to be established, *it will soon be sure to manifest itself.* The *non*-occurrence of such an exception against a supposed law is a far stronger argument than the occurrence of hundreds of instances in its favour: and this consideration probably operates far more strongly with most minds than any abstract principle of conviction.

No intuitive perception of external truth.

If there be any force in what has been advanced, then, instead of any primary or inherent principle,—any original element of the mind, enabling it to see the outward world blindfold,—any intuitive internal power to create external facts, any authority derived solely from the interior resources of pure reason to show us physical and material things without reference to the senses, or the like,—the simple analysis of the case would lead us to the more sober belief that the source of inductive certainty, that certainty beyond the mere limits of sense, that superstructure larger than any foundation of facts, is accounted for by natural and acknowledged processes.

The case resolvable into the power of abstraction.

It arises in the first instance out of the power of *abstraction*, acting with unconscious force and powerful rapidity, by whose aid the mind creates what are indeed new conceptions, yet formed only out of materials already furnished, and this *not by addition, but by subtraction* of properties and particulars.

Mainly aided by perception of analogy.

Above all, the process derives its whole force from the discovery and acceptance of sound and well-framed *analogies*, or, as I have elsewhere said, THE

Soul of Induction is Analogy; and higher, more efficacious, and more enduring, as the analogies adopted are more strictly accordant with the real harmonies of nature.

Application of mathematical reasoning.

The application of a higher reasoning to the mere facts of observation which essentially constitutes *science* throughout a large extent of physical research, is mainly effected by the application of those systems of abstract and necessary mathematical truth which have been independently deduced from abstractions respecting quantity in its several species (themselves derived not less originally from experiences of sensible extension, division, and numeration), whence spring quantitative laws and mathematical theories, which confer on the inductive results, whenever they can be applied, a character of increasing certainty and power arising from the higher capacity for generalisation. Thus the two systems react on each other, and we are often enabled to carry on our views, and predict results to which no mere extension of observation could have conducted us.

The process of inductive generalisation indeed

becomes infinitely more rapid, decisive, and well-founded, when pursued in connection with the deductive method. The application of mathematical formulas, if found to apply to the subject, not only leads with greater readiness to general laws, but carries with it a powerful presumption in many cases that it is really the exponent of some actual and higher natural analogy which we could never have collected from any mere observation of facts.

Correspondence of mathematical and physical truths.

Such instances are, indeed, constantly occurring in various degrees; but, in some particularly striking cases, have evinced, to a singular extent, the correspondence between the real, but as yet unknown, laws of nature, and the abstract creations of mathematical conception: as in the well-known instances of the change in polarisation predicted by Fresnel from the mere interpretation of an algebraic symbol, and the fact of conical and cylindrical refraction anticipated from the mathematical theory by Sir William Hamilton.

Asserted *à priori* evidence of physical truths, examined in several instances.

But this assertion of *à priori* evidence is sometimes made with reference to the primary principles of all natural philosophy—the laws of motion and of equilibrium—whether in solids or fluids. It is alleged, that what is announced as the first law of motion, though it may be attested by constant experience, has yet in itself evidence arising out of the nature of the case beyond all experience.

Laws of motion.

Now, in the first place, I would observe, that the very notions of a body in uniform rectilinear motion, or of forces acting on it, are essentially ideas of experience, and certainly could have no application without reference to the real existence of matter and force.

It may be maintained that the law of inertia—that a body will retain motion communicated to it *after* the direct impulse has ceased—is at least deducible as a consequence from higher first principles; but still those principles are themselves nothing else than mere simple facts, or properties of matter, derived from experience.

It is sometimes alleged that, to assert that a body, left to itself, will go on in uniform rectilinear motion *for ever*, is presumptuously to assert what no *experience* can ever justify; and, therefore, if admitted at all, can only be received as an intellectual truth derived from *à priori* principles. But such perplexity would be removed if we only put the proposition thus: a body in motion, &c., *must* EITHER go on for ever, OR its motion must be changed or stopped; but whatever changes, stops, or retards it, is a new force acting upon it, and the question is then reduced to an examination of the action of that force.

Equilibrium.

Again, it has been sometimes asserted, that the first principle of equilibrium—the foundation of the doctrine of the lever—is *axiomatic* or self-evident.

Yet, without going further, it is obvious that the very idea must imply at least the existence of matter, capable of being acted upon by such a force as gravity through the intervention of something material corresponding to the inflexible straight line of theory;—ideas which can only have been obtained ultimately from experience. When some such principles have been adopted, we can then, and then only, by strict deductive reasoning from them, arrive at the theorem of the lever, which we find confirmed by experiment.*

Equal pressure of fluids.

In like manner, it has been maintained that the

* See my "Essay on the Laws of Motion," and "Essay on Necessary and Contingent Truth;" Ashmolean Memoirs: Oxford, 1837, 1849.

first principle of all hydrostatics, *the equal pressure of fluids,* is not derived from experience, but that the mind can pronounce on its *à priori* certainty. Undoubtedly the mind can infer deductively this great law of fluids, as a *necessary consequence from certain other assumptions*, that is, when certain, yet more elementary properties of fluids are known, and taken as the basis of the science, but not otherwise.

The ulterior principles to which the nature of fluids may be reduced, may have been differently viewed and traced upwards to more or less simple elements by different philosophers, but all have adopted, and must adopt, at the outset, *some primary physical* fact or property to start from. The more simple and general the property referred to, the more satisfactory and complete is the reasoning; and it is the main point in such an enquiry to determine *what* are the *fewest* and *simplest* principles we can assume, in proving these first properties and laws. Still, the ultimate principle, however simple, and however far back it may be traced, can of necessity be nothing else than some *physical* fact the result of universal observation; such as must

A consequence from the nature of fluids.

be even the very *existence* of fluids, and without which no reasoning of the kind could be applicable.

Abstract theory may be conceived.

It is, indeed, quite conceivable that a reasoning being, who had never seen a fluid, might imagine and create theoretically the conception of such a substance, and might reason mathematically on its properties, such as would follow by strict deduction from the constitution thus assigned to it; but this would not apply to anything in nature until it were shown by experience that these properties were really manifested in some substance to which the theoretical notion might be referred.

But physical observation necessary for application.

Example of the undulatory theory.

This is no imaginary case: it actually occurs in the speculations pursued by so many philosophers on an imagined æthereal medium. From the assumed nature of such a purely hypothetical medium, a supposed assemblage of imaginary molecules, acted on by attractive and repulsive forces and liable to agitations from without, by mathematical reasoning the whole of the refined and complicated theory of undulations has been deduced; which, so far, might for ever remain a barren but most beautiful mathematical creation. Independent observation gives us no evidence of the existence of such a medium, and the theory is in no way founded on experience.

Unapplied till optical facts are introduced.

When, however, by the aid of the *eye*, the phenomena of optics present themselves, we find a vast range of such phenomena which admit of a complete explanation on the assumption of this hypothesis: here, for the first time in the inquiry, a reference to anything experimental or sensible comes in. That it must come in somewhere is clear; yet it would be absurd and untrue to say that such theoretical reasoning alone can give any *à priori* certainty to the optical facts or laws to which it is applied, which must after all have been first founded on some small basis of observation. Nevertheless, such applications of mathematics confer *the highest presumption*, little or at all short of certainty, for generalising conclusions actually observed to be true only in one or two instances.

Inverse square of the distance.

To take, perhaps, the strongest instance which has been adduced. The law of force or intensity varying as the *inverse square of the distances*, it is alleged, and doubtless with truth, is a conception of pure reason (so far as any mathematical conception is so) from abstract geometrical considerations, which must hold good in any kind of supposed *emanation*, radiating equally in all directions from a centre, and undergoing no change of condition, excepting that due to distance only.

But though these geometrical ideas throughout may be pure creations of the mind, yet the idea of any such emanation of actual force, however abstract, must have been derived from some ideas of experience, and certainly can apply to nothing in nature without reference to such sensible ideas.

Equal areas proved abstractedly.

Again: to take what is almost an equally striking instance,—the law of equal areas.

It is undeniably a pure result of reason that a metaphysical point revolving about another metaphysical point by virtue of an impulse conspiring with a centripetal force tending to that point, varying according to any law whatsoever, must describe areas proportional to the times.

Inapplicable without some ideas from experience.

But how do we get the idea of a centripetal, or of an impulsive force, unless, at least in the first instance, by abstraction from observed facts? Wherever these forces exist in nature, we reason deductively to the conclusion of a description of equal areas, and we find it confirmed by observation.

But this is a very different thing from gaining independent *à priori* evidence for physical facts.

Paradoxes.

From expressions sometimes used, it would even seem that additional force is supposed to be given to the argument for abstract conception as the ground of physical truth, by the allegation, that some of those primary abstract physical principles to which we have referred are even *opposed* to what mere sense and experience would naturally expect, and must therefore be ascribed to a higher faculty of internal reason; and this, it is also alleged, is practically evinced by the circumstance that such truths are appropriately termed *paradoxes;* as, *e. g.*, the primary property of fluids has led to what is called the "hydrostatic paradox." But this is not owing to anything in the abstract nature of the reasoning.

All truths paradoxes to prepossessed minds.

What does a *paradox* really imply? Any new truth, even a mere matter of observation, is a paradox in popular estimation, if it contradict a received prejudice. The existence of Jupiter's satellites, and the fall of unequal weights in the same time, were paradoxes when announced by Galileo* to the Ari-

* Vignette at the beginning.

stotelians of his day. Yet these were facts of *observation.*

The Aristotelians had held that motion can only be caused by something in *contact* with the body moved; hence the law of inertia was a *paradox* when first asserted; and so, indeed, it continued to be long afterwards even to the Copernicans, as appears from the difficulty they felt in accounting for the continual keeping up of the planetary motions. The application of abstract reasoning in such cases tends, in fact, to *remove* and *explain* the paradox, not to create it. The startling nature of the assertion, therefore, is no proof of its being derived from any intuition superior to sense.

Deductive proof only from physical principles more or less remote.

The question between the *inductive* and the *deductive* process is merely a question of *degree:* in some cases the abstract part of the process may be longer, and its origin more remote from material facts—in others less so. The very same conclusion may often be arrived at by several distinct trains of reasoning, setting out from principles of lower or of higher degrees of abstraction; but there must always be, *somewhere* in the process, a recurrence to sensible experience.

Example. Theory of pendulum.

For instance, without any knowledge of mathematical theories, we might discover *experimentally* and empirically the laws of the motion of the *pendulum;* and so might regard them as *mere facts of observation.*

But, again: if we knew in the first instance, by experimental trial, the law of *falling bodies,* we could deduce mathematically what *must* be the law of the pendulum,—that is, it is a necessary consequence in reason from a simpler mechanical truth, provided that reason be first furnished with that simpler truth.

But, once more: the law of *falling bodies* itself is a necessary consequence of still simpler principles: if we knew, experimentally, the nature of terrestrial gravitation, we might deduce, by pure reasoning, the law, that the spaces described under its influence by bodies falling near the surface of the earth must be proportional to the *squares of the times;* and thence deduce the laws of the pendulum.

But even, still further: if we investigated, *on pure theory,* the effects of *a constant force,* we should deduce the same law for bodies moving from a state of rest under its influence, and this would apply directly to the deduction of the laws of a body constituted as a pendulum under its influence; and hence the laws of the pendulum, as actually moving under the influence of terrestrial gravitation, might be said to be deduced *from pure theory* and the abstract idea of a constant force.

But the real application of such reasoning essentially involves the actual existence in nature of such a force as that of gravity, which can only be derived from observation.

Foucault's pendulum experiment.

If the deviation of Foucault's pendulum had been originally a mere matter of *observation*, it would have been long before experiment would have arrived at the solution. Many would have been the hypotheses of peculiar magnetic, electric, or other causes, for the observed deviation.

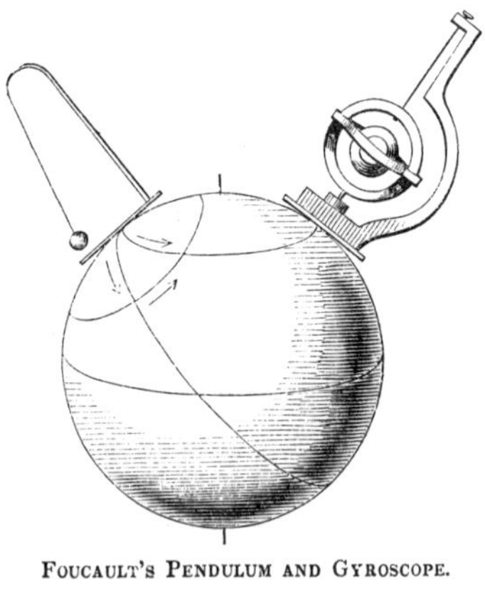

Foucault's Pendulum and Gyroscope.

It was only from a just mathematical conception of the *resolution of the rotatory motion* of any point of the earth's surface into two,—one *round that point*, the other at right angles to the former, and which would not affect the plane of the pendulum's vibration, while the former would,—that M. Foucault foresaw the result. But this was *not à priori* reasoning disclosing a physical fact; it was simply reasoning deductively from a known fact to a consequence; when the reasoning being logical, that consequence could not but be true and be confirmed by observation.

Paradoxes of the gyroscope.

Yet more astonishingly paradoxical are the effects exhibited by means of the gyroscope, which seem to subvert all the acknowledged principles of equilibrium. To mention one only: a wheel loaded round its circumference, in rapid rotation at one end of a horizontal axis, having the other end merely resting on a pivot, is *supported on that pivot alone against gravity*, the whole at the same time revolving round the pivot.

Scarcely less remarkable is the application of this instrument by M. Foucault to another manifestation of the earth's rotation:—the wheel retaining its original plane of rotation, which therefore *apparently* deviates with the rotation of the earth.

It is probable that any person, even of considerable mechanical and experimental knowledge, seeing the action of the gyroscope for the first time, would be much puzzled to account for it, as, in fact, several persons have been; and if he set about investigating it experimentally and *inductively*, might be long before he traced any law or connected it with any principle, so as to reconcile it with the established doctrine of equilibrium.

If, however, he set out with a mathematical knowledge of the principle of the "composition of rotatory motion," and proceeded *deductively*, the explanation is easy, and its relation to a number of other important cases readily manifest. Yet the application of this mathematical theory requires the idea of a *material* body in rotation.

Inclined plane.

The ancients, notwithstanding all their refined geometry and spirit of abstract speculation, were unable to advance to the solution of the case of oblique equilibrium, or the inclined plane; and this is clearly a case where, if anywhere, *à priori* principles would have availed. But it was not until Stevin reasoned, *not* upon any abstruse axioms, but on simple mechanical considerations, that the demonstration was discovered. The solution was effected by reasoning *deductively;* but it was deduction from principles obtained on a primary physical or experimental property of matter.

Discovery of magneto-electricity.

A highly instructive instance of the application of an abstract principle to physical discovery may be found in the way in which Faraday reasoned to the discovery of magneto-electricity, which I cannot describe better or more briefly than in the words of Mr. Grove*: —

"The discovery of Œrsted, by which electricity was made a source of magnetism, soon led philosophers to seek a converse effect; that is, to educe electricity from a permanent magnet. Had these experimentalists succeeded in their expectations of making a stationary magnet a means of electric currents, they would have realised the ancient dreams of perpetual motion — they would have converted statics into dynamics — they would have produced power without expenditure; in other words, they would have become creators. They failed, and Faraday saw their error: he proved that to obtain electricity from magnetism, it was necessary to super-

* Lecture on Progress of Science; London Institution, 1842, p. 20.

add to this latter, motion; that magnets, while in motion, induced electricity in contiguous conductors; and that the direction of such electric currents was tangential to the polar direction of the magnet—that as dynamic electricity may be made the source of magnetism and motion, so magnetism conjoined with motion may be made the source of electricity. Hence originates the science of magneto-electricity, the true converse of electro-magnetism."

The application of mathematical reasoning to physical inquiries may sometimes, at every step, exhibit something corresponding to an actual step in the mechanical process, and thus capable of a physical interpretation: such is often the case in the older geometrical investigations. But in the prevalent applications of the modern analysis there is no correspondence of this kind; the original conditions being once put into an equation, we resign ourselves to mere symbolical operations, which have individually no reference to any physical ideas, till we find ourselves landed as it were on the platform of a conclusion which marvellously harmonises with experimental results.

Yet these and the like instances are not at all

cases of an *à priori* discovery of physical truth; they are instances of a train of logical reasoning proceeding from some first principle derived from remote physical abstractions till it arrives at a conclusion which coincides with some other observed law having no other perceptible connexion with the first principle; or leads the philosopher to expect such a result; which, on trial, is found to be the fact.

Conclusion.

No really *à priori* proof of physical truth.

Thus a simple analysis of the actual train of argument tends to dispel the mystification and confusion which have sometimes arisen on the subject of abstract reasoning applied to physical subjects. Pure reason out of its own resources may, indeed, create theories apart from all observation of nature; but to make them applicable to anything in nature, such creations of the mind must necessarily and universally involve *some* small assumption of *material properties* or mechanical conditions; which can only be in some form or another ultimately derived from observation: what is borrowed may be *very little*, but it must be *something;* and it is a point of interesting research to the philosopher to endeavour to ascend to the fewest and simplest possible of such first principles.

Necessary truths only necessary consequences.

A confusion of ideas is sometimes introduced by the use of the term "necessary" dependence or "mechanical necessity;" as if it were a *blind or fated necessity;* but what we mean is, a *necessity of reason* or *logical sequence.* It is evinced by the dependence of a series of ideas deductively followed out; which are also found to accord in their result with natural facts and more comprehensive laws.

Speculative ideas of Œrsted.

The subject here discussed, is beautifully illustrated by the philosophical views broached in a posthumous work*, which has so fitly and honourably crowned the labours of the great Œrsted, and added a new claim to our admiration of his genius. In those essays he maintains repeatedly the proposition that "the laws of nature are the same as the thoughts within us;" "the laws of motion are such as are required by our understanding;"† "the law of the inverse square of the distance is a conception of reason;" and several like instances: all which I should fully admit, *subject* to the qualification above suggested and understood in the sense

* "The Soul in Nature," translated by the Misses Horner. London, 1852.

† See especially pp. 10. 36. 93.

which it implies—that the connexion and dependence of the facts in nature accords with the connexion and dependence in our reason, *provided* we set out from some more or less simple principle originally *derived from observation*, whence we advance by abstract reasoning to a conclusion, which, however remote from the physical point whence we started, is found to accord with natural facts, and to be a general law of nature. In this sense I have above considered some of the cases just referred to; and others adduced by Œrsted are more obviously of the same kind; such as the lesser planetary and lunar perturbations, too small for observation alone to detect, yet indicated by theory; the identity of lightning and electricity; the discovery of the metallic bases of the earth: all anticipated by theory. To which might be added Œrsted's own grand discovery of ELECTRO-MAGNETISM, and that of the planet Neptune in our own day. But these cases are, after all, not precisely in point to the original question, since here the starting-point was obviously previous *inductive* knowledge.

These distinctions are important to the fundamental analysis of our reasonings on which we ad- **Accordance of reason and nature.**

vance legitimately to those broader ulterior reflexions on which Œrsted enlarges, and which are the same to which the whole of the present inquiry points.

Œrsted has well remarked that it is a common error to imagine matter something constant and invariable. But the permanence and invariability of nature are not found in its individual parts, which are all undergoing perpetual changes. The invariable, he argues, is found only in the abstract nature of things: "nothing is invariable in nature but *laws* which may be called the thoughts of nature."*

Natural combinations (Œrsted observes) which appear accidental are not really so. "All effects obey natural laws; these laws stand in the same necessary connexion as one axiom in reason to another: that this combination is precisely a combination of reason we learn from this, that by reason we are enabled to deduce one law of nature from the other, and by the known laws to discover new and unknown ones. Innumerable as are the effects determined by natural laws in every object in nature, however insignificant it may be, I deeply feel an

* The Soul in Nature, p. 23.

unfathomable reason within them, of which I can only comprehend by fragments an incalculably small part. In short, nature is to me the revelation of an endless living and acting reason."

"If the laws of reason did not exist in nature, we should vainly attempt to force them upon her: if the laws of nature did not exist in our reason, we should not be able to comprehend them." *

And on the whole, "we find an agreement between our reason and works which our reason did not produce." . . . "All existence is a dominion of reason." "The laws of nature are laws of reason," and "altogether form an endless unity of reason," . . . "one and the same throughout the universe." †

* The Soul in Nature, p. 18.

† Ib. 12. 16. 87. 92. 377.

§ II.—THE UNITY OF SCIENCES.

Tendency of sciences towards common principles.

All branches of inductive science continually tend more and more towards a grand *unity of principle*. We perceive this to a partial and limited extent in every lesser advance of discovery: in proportion as new facts accumulate and become embarrassing from their multiplicity, sooner or later some happy advance in generalisation is always found to occur by which they are simplified and reduced to some single principle, connecting them at the same time with other classes of phenomena.

Gradual progress of generalisation and union of sciences.

In the science of the ancients (exact as it was in some limited departments, each within itself) all branches were isolated and disconnected: and all physical principles and causes were supposed of separate and even conflicting kinds.

First advances.

All the first great modern advances were directed towards *combining* and *uniting* branches hitherto distinct, and tended to evince a *unity* of idea and principle pervading them. The first discoveries pointed to the identification of the celestial motions with terrestrial; of astronomy with mechanics; of the fall of an apple with the motion of the moon; of the horror of a vacuum with the laws of equilibrium: as later discoveries have identified magnetic and electric currents, and connected sound, heat, and light with the mechanism of waves; and, again, the resulting effects of heat with dynamical force.

Later physical discoveries.

Faraday's generalisations.

Of the tendency and progress of discovery towards a coalition and combination of different trains of research, perhaps we can nowhere find more striking instances than in the multitudinous researches—and every *research* terminates in a *discovery*—of FARADAY. The peculiar character of high generalisation which results out of an apparently immensely complicated mass of small details, is perhaps one of the most striking features of this wonderful series of investigations. It is impossible here to do more than select one or two instances.

Magnetism and diamagnetism.

Few generalisations of a more striking character have ever been announced than that of the *magnetic properties of all matter*, evinced in the classification of all substances under two species, magnetic and diamagnetic, and these characterised respectively by the properties of attraction and repulsion.

Action of magnetism on light.

But in this *union of relation* between magnetic and all other matter, there was to be disclosed a yet more striking instance *of bringing together remotely separated* kinds of physical action under a common law, in the action of *magnetism on light.*

What could be a more singular and striking *identification* of properties in cases *apparently* the most *remote* from each other than the production of *rotatory polarisation* in light passing through *quartz* and some other substances, and in passing through ordinary *transparent media when placed in the line of intense magnetic force?*

Or to go back to an earlier discovery:

Definite electrolysis.

Grand indeed was the conception of the principle of the relations of *chemical* to *electric* action, partially illustrated in theories of Fabroni and Wollaston, but first announced in all its generality by Davy: thus bringing into *close relation and unity* two such great modifications of physical power. Equally important, though apparently remote from either of the last was the principle of *definite proportions in atomic combinations* disclosed by Dalton.

These two comprehensive generalisations, each equally wonderful in itself, yet seemingly unconnected, it was reserved for the penetrating genius of Faraday to place in intimate connection and to *unite in a still higher bond of generality.* No single discovery perhaps could be cited of higher intrinsic value than the disclosure of the great principle of DEFINITE ELECTROLYSIS: but the high philosophical character of this discovery is enhanced the more specially in that it *combines in a principle of unity* the mathematical *law of definite proportions* in *chemical combinations* with the preservation of the same identical *numerical relations* in *electrolytic action:* and thus uniting both in intimate relation with the fundamental conception of *atomic composition.*

Systematic division of sciences.

As we look to the larger divisions of the sciences, and the successive wider generalisations which they imply, the same tendency to unity is continually though more slowly manifested. And thus, even where it does not yet appear, we cannot doubt that this is the legitimate and ultimate direction and tendency, however remote, of all scientific progress. But in treating of the sciences systematically, it is necessary to adopt some principles of classification and arrangement. Here some division is rendered necessary for this particular object; but it ought to be carefully borne in mind that it should in no way really interfere with the increasing conviction of a real unity of principle pervading all branches.

Some views tending to isolate sciences.

It is a reversal of the order of inductive advance to endeavour to isolate each department of science, and to place it on a separate base, by a theory which would assign to each branch certain real differences of principle and peculiar fundamental ideas essentially characterising it. If such a distinction were made out, it could be but a temporary and provisional ground of classification, in time to be superseded by a reduction to a higher common principle.

Mechanical force.

It is no doubt true, that the highest, the most perfect, and satisfactory assignment of physical causes is effected when the phenomena can be analysed into *mechanical* laws. But the reason of this lies in no mysterious connection of mechanics, as such, with the idea of causation, but merely in this, that the conditions of purely mechanical reasoning are so perfectly elementary in their nature, and so entirely free from all admixture of ambiguous or doubtful conditions, that we can directly investigate them with a simplicity differing in nothing from that of primary geometry, and thus attain the most perfectly satisfactory explanation, when everything is reduced to simple consequences of mechanical equilibrium or the composition of forces.

Progress of all sciences towards the idea of mechanical force and motion.

In other branches it is clear that just in proportion as we can succeed in reducing the phenomena from obscure and apparently mysterious modes of action to these simple and intelligible cases of force and motion, in the same proportion we bring those branches into the domain of exact science, and break down the line of demarcation which hitherto seemed to separate them.

Common grounds of statics and dynamics.

The sciences of statics and dynamics, of equilibrium and motion, have been represented by some writers as based on inherently distinct principles: but it is at once a more satisfactory, and as I believe a more true, view which connects them by the consideration that the simplest cases of *equilibrium* or *rest* cannot fully be demonstrated without an explicit or tacit reference to the idea of *motion* *: which thus far helps the more general consideration of the ultimate unity of all sciences.

The explanation of the precession of equinoxes (the same in substance as that of Newton, more circuitously followed out) by the direct application of the composition of rotatory motion announced by Frisi, and imitated by the rotatory apparatus of Atkinson and Bonenberger, exhibits a peculiarly striking exemplification of unity of principle in passing from such phenomena, vast in their relations both to space and time, to the identical cases pre-

* See my "Essay on Necessary and Contingent Truth," Ashmolean Memoirs: Oxford, 1837.

sented in the deviations of rotatory projectiles *, the cases of spinning tops in stable and in unstable equilibrium, and the various paradoxical effects produced by the gyroscope: all, however diverse, direct consequences of one simple law.†

Polarity reduced to resolution of motion.

The idea of "polarity," to which such mysterious importance has been attached, has been sometimes imagined to involve some essential peculiarity supplying an appropriate characteristic conception to mark a distinct class of physical phenomena. But this once marvellous notion, in the instance of *light*, has *been reduced to a simple case of resolution of motion;* and there can be as little doubt that the progress of inductive generalisation, and the application of mathematical principles, will, sooner or later, reduce other instances, at present provisionally designated by the same name, to equally simple modes of action.

Dynamical theory of waves.

And with respect to the phenomena of optics generally, how completely remote do they appear

* See a Memoir by Prof. Magnus, translated in Taylor's Foreign Mem., N.S. pt. iii. p. 210.

† The Vignette at the head of this Section represents the apparatus as constructed on a large scale for lecture illustration at the Royal Polytechnic Institution.

from all notions of mechanical force? yet, by the mathematical labours of Fresnel, Cauchy, and others, these seemingly remote appearances have been connected with a recondite theory of pure dynamics; which, followed out through a complicated train of deduction, ends in reducing nearly all these phenomena to the results of certain minute *motions*, subsisting and excited among a system of molecules acted on by attractive and repulsive forces, and subject to external agitation.

Electro-magnetic currents.

So, again, when electric and magnetic action were reduced to systems of currents by the researches of the numerous and distinguished co-operators, in following out the great principle disclosed by Œrsted, there was a direct approach to ideas of *motions in definite directions*, which supply the abstract indications of force; and though the subject has even yet been but imperfectly followed out, we perceive the direction it is taking, and must eventually take, towards satisfactory explanation, in a reduction to simple dynamical principles.

Supposed case of interference in galvanic action.

One of the most remarkable approaches (as yet quite in obscurity) which has been made towards a connexion in principle between two branches of science apparently remote, is that of a peculiar action of a galvanic current exhibiting all the marks of a case of INTERFERENCE, in the experiments of Mr. Grove.* If this should be followed out by a more close analysis, so as to show a real action of the kind, the analogy of galvanic action with *a system of vibrations* of a fluid analogous to the luminiferous æther as its cause, would open the way to a generalisation of the highest and most valuable kind. And further, it may not be altogether inconceivable that two sets of such vibrations, which, by superposition, give rise to elliptic vibrations, may be connected with the formation of currents running round the wire, by which so many of the phenomena are represented.

Molecular forces.

Again: to insist on an essential scientific distinction between *molecular* forces and those acting on matter in larger *masses*, as the characteristic basis of a peculiar science, tends to isolate this branch from ordinary dynamics, to which we should rather seek to assimilate it.

In the same way the broader distinction between

* Phil. Trans. 1852. Part I.

Chemical forces.

mechanical and *chemical* action tends equally to break up the idea of that essential and fundamental unity which the philosopher is persuaded must really subsist between these invisible actions of atoms on atoms, and those only more obvious, because on a larger scale, of worlds on worlds.

The distinction of *molecular* forces, there can be no doubt, marks merely a *present* line of demarcation from ordinary mechanical forces, which will at some future time be effectually broken down, and the two classes reduced to one higher genus. *Chemical* action, again, we may be assured, differs from *mechanical* only in our existing state of ignorance; but they will doubtless at some period be assimilated by the discovery of a common principle of equilibrium and its disturbance. Even in the present state of our knowledge, molecular forces have been shown with great probability to be reducible to a common theoretical expression with that of gravitation in the speculations of Boscovich and Mossotti.

Cosmical forces.

Again, the mode of aggregation of many of the stellar clusters, as described by recent observation, is regarded by some very eminent philosophers as evincing the action of forces of a peculiar kind different from those of gravitation. Should this prove to be the case, it would in no way derogate from the universality of *some law* of aggregation of matter, that a different species of law may prevail in those vast distant portions of the universe, which, when it shall have been investigated, may prove a more comprehensive kind of force, of which gravitation is but one form or modification.

Outstanding exceptions remain for analysis.

But if any such apparently outstanding exceptional case were fully made out rightly to claim the title of involving an entirely new principle, still the inductive method would only mark out that principle as a legitimate subject of future analysis; and we might be assured that in the successful course of such analysis at some future period, either this new principle must fall under some already recognised principles, or those recognised principles must fall under it.

Distinctions temporary and provisional.

There may no doubt be a practical convenience in retaining some distinctions of this kind to *preserve arrangement* in our subjects; but to attempt to fix them as essential foundations of real philosophical distinctions, seems to be reversing the proper order of inductive inquiry. *Provisional* and *temporary* distinctions for classification, indeed, we may with convenience and advantage often make between different branches of science in regard to the modes of reasoning and nature of the leading ideas appropriate to them; but it is essential to remember that these distinctions are *only* provisional.

Alleged exception as to geology.

But in contemplating the unity of sciences, an exception has been alleged in reference to GEOLOGY. The entire relation in which it stands to other branches of *inductive* science, and even its *inductive character* altogether, has been sometimes disparaged. Comte has most unaccountably denied it any place whatever in the scheme of "positive philosophy," and possibly some hypotheses which have continued to be occasionally indulged in, in connection with that science, might not unnaturally have influenced him in entertaining a prejudice against it.

Not real: geology an inductive science.

Yet this science, when rightly pursued, is eminently *inductive.* From its very nature it combines the resources of a variety of other sciences; dynamical, hydrostatical, chemical, and especially physiological, and being thus entirely dependent on these other branches of inductive philosophy, itself acquires a perfectly strict *inductive* character.

When, at the present day, it exhibits to us, preserved in their stony sepulchres, the successive varieties of organised structures, as they lived and moved in the same world, subject to the same immutable laws, mechanical, optical, and physical, uninterruptedly in operation through all the incalculably vast periods of past time, it is an entire departure from all just appreciation of the unity of science and of nature to imagine that any essentially different laws of vitality then prevailed, or that the changes in organised life thus brought to light were governed by any totally different series of causes from those now in operation of a peculiar and mysterious kind.

No really different causes referred to.

Yet some seem to have supposed that the reasoning of geology ought to rest on something distinct from that of the experimental sciences,—inasmuch as it refers to events which have so long since passed away, and which we cannot recall for examination, while the very terms "*palæozoic*" and "*palætiology*," might seem to insinuate that we are concerned with an order of causes belonging to the past, different from those now in action,—a distinction just as unphilosophical as that of the peripatetics, who drew a distinction between "natural and violent" motion,

The evidence of geology not different in nature from that of other sciences.

and ascribed the terrestrial motions and the celestial to distinct causes.

Induction has no reference to distinction of *past* or *present;* if phenomena have been locked up for ages, yet, when once thrown open to us, they become objects of the same kind of investigation as those occurring at the present day. The investigation and restoration of the remains of a Saurian imbedded millions of ages ago, is an operation of precisely the same kind as the *post-mortem* examination of the subject of yesterday.

Uniformity of nature in time and space.

The inductive philosopher is convinced that the universal subordination of causes must hold good equally in *time* as in *space;* that as there is no *region,* however distant, in which physical laws do not apply, or in which, if as yet unknown, we are not fully warranted in feeling an assurance that they must apply; so in *time* there is no *period,* however remote, at which we can legitimately imagine the chain of physical causation to be broken, and to give place to disconnected influences of a wholly different kind.

Geology approximating to an

More recently, the investigations of Mr. Hopkins have tended to connect geology even with dynamics and mathematical laws, and thus to establish its relation, not merely to the *inductive*, but even to the *exact* sciences: not that that name implies any real difference in nature, but merely marks the *degree* of perfection to which any branch of science has attained. If, then, from the examination of phenomena *actually existing*, and going on around us, we turn to the past, the rules and principles of inductive investigation will apply with equal force and propriety to phenomena which teach us the successive and gradual changes which the crust of the globe has undergone, and lead us to trace them as far back as we can towards its origin.

exact science.

The great principle which forms the basis of all inductive geology — the analogy of existing causes in explaining past changes must, however, be distinctly understood, and, in fact, is so interpreted by its best advocates, not merely as restricted literally to those identical natural operations which *we see going on*, AND COMPLETED, daily before our eyes within the limited moment of time to which our observation extends.

Influence of time admitted.

It would not fully vindicate its own power, if it did not include in the general analogy the influence of some elements *incapable, from their nature, of direct verification from our own experience,* such as are due to the INFLUENCE OF TIME, especially of *unlimited periods of time;* and in illustration of this idea we are reminded of some changes even in more limited periods, which, though in their nature and results simply chemical, are yet *such as cannot be,* or at least *have not been, produced in our laboratories.* We may take as instances the formation of *coal* and of *diamond;* while on a grander scale we are under the necessity of acknowledging the long series of changes which must have accompanied the gradual cooling of the earth, an unavoidable inference from the fact of existing central heat.

Illustration from formation of coal and diamond.

Misconception of past duration.

Real inductive principles thus tend to reduce to order those phenomena which have appeared to some to present so much more strongly marked vicissitudes only because we are apt to crowd the events together in the long perspective, and measure them too much according to our confined ideas of duration.

Theory of convulsions.

In speculations on changes where, it is alleged, all applications of known causes fail, it has been the favourite resource with some to appeal to mysterious revolutions and occult operations of a kind ill-explained, and even supposed to be inscrutable to our faculties, but thus the better calculated to dazzle the many with their imposing pretensions.

Of an uninductive character.

But in the spirit of true induction we have no right to imagine that any of the events or changes of past epochs, however apparently inexplicable, can be rationally set down as events of a *different kind and order* from those now going on, or as interruptions of the settled order of natural causes.

Uniformitarianism and catastrophism.

Difference of opinion indeed may subsist as to the *greater or less frequency* or *intensity* of volcanic action, of fractures and dislocations, of variations in climate, of changes of condition due to the cooling of the terrestrial nucleus, or the like, in past epochs. But these, while they are on all hands allowed to be fair and legitimate topics of philosophical debate and inductive inquiry, would be most unduly exaggerated if supposed to mark any such real or fundamental difference in principle as to constitute two really distinct geological schools. They are questions merely of *degree*, not of *kind* or of *principle*.

Yet, in the language often used, the " uniformitarian" view would seem to be represented as an hypothesis to be fairly weighed against another antagonistic "catastrophic" theory. If the terms are to be understood with any such difference of sense as that thus implied, I conceive it appears that the two theories respectively occupy totally different grounds.

The "uniformity" principle would mean simply the proper extension of inductive analogy and the law of continuity, even if not yet sufficiently substantiated in detail in each particular instance; while the "catastrophic" hypothesis seems of an essentially uninductive nature, and appeals to ideas remote from true analogies, confessedly resorted to on the very plea of the failure of explanation by natural causes.

But, in such cases, the evidence of a violation of the uniformity of nature is purely *negative:* with all analogy against the reality of the exceptions, they *can be such only to our present ignorance:* the apparent anomaly is but a part of a more comprehensive law, ill understood; — a modification of its continuous action in reality equally regular, though not as yet fully made out or reduced to law. Geology thus kept pure from the introduction of fanciful and unphilosophical hypotheses eminently conforms to the type of unity which binds together the whole range of inductive science.

Revolutions in science only progressive.

The unity of sciences is not impaired, but insured and promoted, by those *mutations* which any of its branches may seem to have undergone. All real science is in a state of perpetual change. These changes have now and then been fundamental and revolutionary, and similar fluctuations are perpetually going on in lesser details. But this in no way makes science itself unstable or fluctuating. The change is always of one character, and that no other than the very nature of the inductive philosophy requires: a change from anomaly to regularity, from heterogeneity to analogy, from confusion to order, from interruption to continuity, from artificial dogmatism to the simplicity of nature.

Discoveries superseded only by greater improvements.

Every branch of science approaches perfection and stability as it more fully approaches to and realises the grand principle of *unity*. It is the test of the real advance of discovery to exhibit a progressively increasing conformity to these great principles: an advance which will not require a retreat,—the erection of a structure which will not require re-modelling.

Every philosophic research or conclusion, at present of the highest importance, must expect to be reduced to a subordinate place: every method now most justly esteemed must look to be superseded by greater improvements: but nothing will deprive such really great discoveries of their place in the page of history—their lustre will but be increased by the brilliancy of newer results, to which they were the necessary preliminaries.

Such mutations are sometimes made a topic of reproach, but only by those who are hostile to science from entire ignorance of its principles; they may learn to observe that these changes are *all in one direction:* they are all steps in advance towards a higher and more enduring system—all future progress must be in the same direction; we shall never see a recession from the more natural towards the more mysterious; from the recognition of regulated causes, law and order, in a retrograde course towards arbitrary or fortuitous influences.

Advance from mysticism to reason.

Most sciences had their origin in the clouds of mysticism, and thus occasionally long retain some tincture of it. Astronomy arose out of astrology, chemistry out of alchemy, and geology out of a theological cosmogony. Geology, indeed, being the youngest of the inductive sciences, has naturally in the course of its rapid growth, within a brief period, exhibited more of those changes from mysticism towards rationalism than any other branch. It is but a short time since the whole science consisted of little better than a few detached general facts, connected by arbitrary hypotheses, and conformed to the language of dogmatic belief.

Applied to geology.

With an increasing recognition of true inductive principles, we have witnessed progressive improvements in the philosophic character of the theory and candid retractations of opinions once warmly upheld, chiefly on grounds alien from those of science. Yet these concessions perhaps were made more from the disclosure of a few contradictory facts in particular instances, than from any perception of broader philosophic principles as those which in the first instance ought to have formed the basis of the whole science; and, perhaps, such principles are hardly yet universally recognised in their full force and extent.

Those who continue really to indulge in the visions

Influence of dogmas.

which misled geology in its infancy, the dreams of universal cataclysms, and sudden creations, of a kind wholly remote from physical analogies, and to which it would be wrong to seek to apply physical explanations, so far place themselves out of the pale of the inductive philosophy.

But the influence of such artificial theories we may be assured will in time entirely disappear, and all true cultivators of science will come to regard such distinction of schools in no other sense than as we now speak of Ptolemaists and Copernicans, Cartesians and Newtonians: these anticipations, however, are far from being yet generally realised. Many who smile at the fancies of a Whiston or a Buffon are scarcely less under the dominion of ideas of very kindred origin. Those who disown dogmatic authority to teach the mode of formation of the earth's crust are yet often not exempt from prepossessions equally narrow in speculating on the probable order of creation, the succession of species, or the relations of our globe to other planetary and stellar worlds.

Tendency to principle of unity.

But to minds duly impressed with the great principles of analogy, law and order, all anomalous imaginations derived from sources extraneous to science will disappear. The increasing tendency of all research towards harmony, simplicity, and unity of character, will be recognised as a pledge of its ultimate realisation: and even conjectural hypotheses, confessedly a mere indulgence in philosophical romance, provided it *be* strictly philosophical, will be hailed with satisfaction as helping out the general conception and keeping alive the spirit of analogical inquiry.

Second alleged exception in the science of organisation and life.

But a yet more serious question, of the same kind as that referring to geology, has been raised with respect to the *sciences of organisation and life:* which are sometimes supposed to involve altogether a *new class and order of ideas* of so peculiar a kind that they must stand out as entirely exceptional cases to the general *unity of the sciences.*

Peculiar difficulties of the subject.

Now it will on all hands be allowed that these subjects are as yet but imperfectly understood, and a large range of inquiry connected with them still involved in obscurity. And if from external phenomena we seek to advance to their causes and principles, it is of course most fully admitted that of the ultimate causes of organisation and life we cannot *at present* attain to any satisfactory explanation, or even form any definite conception.

But hence we find many in treating the subject commonly set it down as in its own nature *something essentially mysterious and inscrutable:* as referring to an order of causes altogether distinct, wholly disconnected with those of any branch of physical investigation; as involving functions and operations wholly *sui generis:* and not only that we *cannot* explain them on any merely physical principles, but that we *ought not* to attempt to do so: that they are of an order wholly transcending such inquiries; beyond the power of our faculties to apprehend; and ought to be kept apart, as being indications of a special and mysterious principle which it would be presumptuous and immoral to attempt to inquire into.

Not really mysterious or inscrutable.

Everything doubtless is mysterious till it is made known, but the inductive inquirer will never allow the apparent obscurity of a subject to oppose any barrier to the endeavour to make it clear. Nothing can be more *mysterious* than gravitation; but that does not hinder the philosopher from investigating its laws, or thence, as far as he can, penetrating towards its principle. Electricity and magnetism, thunder and lightning, were perfect mysteries a century ago. Instead of allowing any such prepossessions to paralyse his researches, the inductive philosopher would simply seek in regular order, first to determine the *external* conditions and *laws* of life, themselves as yet far from being well understood. Until these are known, he might reject as premature, or at least regard as wholly conjectural, all attempts to speculate on their higher laws or physical causes: yet not less confidently would he be assured that these more interior causes will one day come to be known; just as surely as the proximate laws will be accurately traced and reduced to that determinate order which undoubtedly in reality pervades them, but of which we have at present only the most imperfect glimpses, yet which, imperfect as they are, are the true openings to the ultimate inductive knowledge of causes and principles.

Proposed hypotheses of the vital principle often fallacious.

There have not been wanting, indeed, attempts at theorising on the subject: various hypotheses have been started as to the nature of the "vital principle," and the question discussed whether life is the result of organisation, or organisation of life. Some have referred to more particular modes of action, such as electric currents flowing through the nervous system, or the like; and have represented animated beings as in fact nothing more than walking galvanic batteries: all these, and many similar theories, *may* be utterly fallacious and erroneous; and the opponents may triumph and revel in the real or supposed refutation of them. But all this in no way affects the conviction of the existence of *some physical principle*, the cause of the vital functions, as yet, indeed, unknown, but which nevertheless will, at some time, become as well determined as the principle of respiration or the circulation of the blood are at present.

Again, though chemical analysis has reduced organised products to determinate elements, yet it is made a matter of no small boast by some, that no chemistry can reproduce an organic substance, or invest that organised substance with life: and eager and loud was the triumph of those who conceived they had refuted the alleged results of Messrs. Crosse and Weekes, and bitter the abuse and ridicule heaped upon them for believing that they had evolved insect-life by galvanism.

But some physical cause of life.

All such experiments *may* indeed be fallacious and premature; and we *may* be as far as possible from at present penetrating the secret of vitality, or the precise mode of its connexion with the bodily structure and the chemical changes elaborated by the various organs. But the truly inductive inquirer can never doubt that there really exists as complete and continuous a relation and connexion of *some kind* between the manifestations of life and the simplest mechanical or chemical laws evinced in the varied actions of the body in which it resides, as there is between the action of any machine and the laws of motion and equilibrium,—the weaving of cloth by a power-loom and the principle of latent heat: and that this connexion and dependence is but one component portion of the vast chain of physical causation whose essential strength lies in its universal continuity, which extends, without interruption, through the entire world of order, and in which a real disruption of one link would be the destruction of the whole.

The principles of inductive science apply to all

All nature subject to law and order.

physical truth and the investigation of all *physical* causes. The laws of order, uniformity, and continuity belong to all parts of the *material* world: and in this order and continuity *animal life* is included. From the lowest mechanical or chemical influences on inorganic matter, there is an unbroken series to the first manifestation of organic changes; and from these again—from the lowest vegetable or zoophyte up to the highest mammalia—there is entirely one continuous progression, its connexion from one term to another being carried on through absolutely insensible degrees and shades of difference.

Humboldt observes,—"All myths about imponderable matters and special vital forces inherent in organised beings, only render views of nature perplexed and indistinct."* It is the unbroken preservation of this continuity which assures us that the nature of the vital principle must be sought for by no occult or mysterious process, but only by the patient application of the same inductive processes by which other physical principles have been and

* Cosmos, 69. transl. 1845.

always continue to be gradually cleared up and elicited; and by the operation of which, we may be assured, this hidden spring of life will, at some time, be disclosed, and brought out to occupy its place in harmony with all the other great principles of the universal cosmos.

Physiology supposed distinct, as founded on final causes.

But there is another plea on which the physiological sciences have been sometimes supposed to stand apart from other branches. It is alleged they are characterised by involving the peculiar and distinctive idea of *organisation*, that is, an idea essentially involving the conception of *design* or *intention*, and have hence been referred to a separate principle called *teleology*.

This, however, appears to me a distinction unfounded in itself, or rather founded on an *incidental* and not on an *essential* distinction, and referring rather to the narrower view of this class of investigations as followed by an older and less advanced school; whereas in their more modern extension, they imply a more enlarged principle, and one closely accordant with the extension of analogy and the unity of science.

It is of course obvious that throughout these

The distinction not tenable.

sciences, perpetual instances of such adaptation of structure to the ends and purposes of life are abundantly manifested; and it is no less evident that they force themselves on the mind with that peculiar, immediate, and irresistible kind of effect which is justly dwelt upon by most writers on the subject, and admitted by all inquirers in such multitudes of convincing examples. On these, however, it is not my object to enlarge here; the present question is as to the precise philosophic analysis of the case with reference to the classification of sciences.

So *rapid* is the mental operation by which the inference of design in these cases flashes upon us, and so *immediate* is the impression, that it may seem almost to *precede*, or at least to go hand in hand with observation, without waiting for formal deduction: so that we may not unnaturally deceive ourselves, and may sometimes mistake it for an intuitive notion, acquired antecedently to the actual examination of organised structures, and may even imagine (as some have even maintained on philosophical grounds) the idea of a purpose, and end and means, is an integral part of our very idea of an organised being. Yet when we analyse our conceptions more strictly, it must be apparent that our very notion of the existence of organised beings must be acquired in the first instance from observation—including the observation of ourselves: and this constitutes so constant and universal a case of experience, that it may well seem an idea whose origin we may set down as contemporary with our earliest exercise of consciousness and thought.

Not essential, but incidental.

It is, however, in strictness, not merely from observation, but by a considerable exercise of inference and deduction, that we can legitimately arrive at the notion that an animal "is intended to live;" it is derived from the study of its organisation; whence we are led to look to the subserviency of its parts to the purposes of life and enjoyment.

The idea which we form in general of an *organised body*, no doubt practically involves that of parts mutually dependent and adapted to each other; but this is an *inference*, and the relation which it establishes is one in no way *essentially* differing, in this respect, from that existing among the component portions of a moving machine, or even of a stationary arch; though certainly differing in the *degree* of complication, as in the higher and more varied and multiplied design and object evinced in their structure. There is no essential distinction *in kind* between our conception of the one or the other. It is true we soon come practically and habitually to include these effects in the complex current idea of an organised being, and are unconsciously and involuntarily led as it were to connect these conditions with the idea of plan and intention, and to assume the relation of these as that of ends and means. But we are here concerned only with the *analysis* of our ideas.

More enlarged views necessary.

That a combination of arrangements, perhaps even complicated ones, which answer a purpose whose practical importance is obvious, and where the relation of one to the other as end and means almost forces itself on the mind the moment we contemplate them, must produce a high conviction of design, is as indisputable as it is invaluable in the high argument of which it forms a part. But such instances arising in the contemplation of *organised* structures do not stand in any way *peculiarly* distinguished *in their nature* from other cases of the like adaptation of means to an end in the wider arrangements of *unorganised* matter.

It may readily be granted as the fact, that several remarkable physiological discoveries have been made in consequence of the habit of looking at final causes in animal structures. But what does this prove? Reduced to its proper place in the philosophical system, the case is simply this: most discoveries in physical science are originally prompted and suggested by some previous *conjecture.* Nothing can be more fruitful in furnishing such conjectures than the habitual recourse to instances of adaptation to an end in organisation already known, whence the enlightened physiologist often receives the most valuable hints, and frames the most probable conjectures as to those which are as yet unknown. The value and force of such conjectures in general depends on the happy preservation of *analogy;* and that analogy is in these cases most likely to be traced in the connected series of means and ends.

Discoveries made by help of final causes; only as hints.

The object is not in this place to enter on the general argument of "final causes:" and in reference to the present subject I will only remark, that the wider extension of physiology by the introduction of the more enlarged and modern principle of "unity of composition," besides its proper claims

Unity of composition the true principle of philosophical physiology.

as the basis of all great and scientific conceptions of such subjects is also remarkable in this respect, that it leads us more directly to recognise the proper place of physiology among the sciences as exhibiting it more clearly in its relations to that *unity* of principle which pervades them all. There is nothing *exclusive* or peculiar in the study of organised bodies; it involves no essentially characteristic idea distinct from other branches of physical investigation, but, like them, tends to the grand conclusion of a reference to common and high principles of unity and harmony of plan and design throughout nature.

Hence unity of physiology with other sciences.

Difficulty as to including man in the series of nature.

But the most difficult, and at the same time the most important question in any theory of this kind, has been raised on the ground of its relation to *the nature of* MAN.

It will, however, hardly be denied that man, *considered in his animal nature alone,* is very little superior to brutes, and in some respects inferior. In the scale of *mere animal organisation,* the difference between the lowest human form and the highest monkey is not greater than between one class of monkey and another. Whatever difference of opinion may have arisen on this subject of a moral and metaphysical kind, yet it is on all hands allowed that man has *to a certain extent* a nature in common with brutes: and we may *avoid all cavil* if we simply assert that man, *in so far as* he partakes in a nature common to brutes, is along with them, in *that respect,* a part of the same scale and system of organised life. *In so far as* his animal nature, functions, and instincts are concerned, they are linked in the same chain of continuity with the order of other material existences.

Distinction of man's nature; animal and higher.

To *what extent* mind and volition, especially in their lower functions, in *man* are different from the corresponding manifestations in inferior animals, is doubtless a very important question of psychology. To draw the line may be difficult or impracticable. Without pretending to determine such a point, we may safely say that, *in so far as* they belong to the animal part of man's constitution, the question as to the nature of such manifestations of intelligence may be a question of *degree,* and may be philosophically treated as connected with other questions of man's physical development, as part of the great scale of natural existence, governed by natural laws as yet very imperfectly known, but fairly subjects of inductive inquiry. The question of an intellectual principle is of a peculiarly *metaphysical* kind, and in no way affects the continuity of man's *physical nature* with the rest of the material order of things.

Man's spiritual nature of a different order of ideas.

But the more important question refers to the further assertion of a distinct *moral and spiritual nature* or principle existing in man, and all the higher relations consequent upon it, which place the nature of man *in this respect* in a category altogether different from that of inferior animals.

Now on this most important point I would only observe one thing in reference to our present subject: the assertion in its very nature and essence refers wholly to a DIFFERENT ORDER OF THINGS, apart from, and transcending, any material ideas whatsoever: hence *it cannot be affected by any considerations or conclusions belonging to the laws of matter or nature.*

Man connected in the natural series.

In a word, man's nature and existence on earth is in nothing of a *peculiar* kind, and in no way violates the *essential unity and continuity* of natural causes:—in regard to man's *animal* nature, because, *so far as that extends,* it wholly belongs to the physical order of things;—in regard to man's *spiritual* nature, because, so far as that is concerned, it is wholly independent of all material things, and is therefore relieved from all possibility of connexion, or collision, with any physical truths, or theories.

Similar distinction as to the past.

As, then, the foregoing consideration refers to the study of the *existing* relations of organised life and of man's nature, *so far as* it belongs to animal existence, so the same principles equally apply to the investigation of its *past* history and origin, so far as we can trace it. We need seek for no more peculiar or occult cause in the one case than in the other.

If we admit that the earth, being still hot internally, must have cooled at its surface, and that this cooling must, in its progress, have caused contortions, dislocations, upheavals of strata; and again, that the waters charged with matter must have deposited it; and that the various crystallised bodies and metallic veins must have been formed during certain stages of these formations,—it is only by parity of reason affirmed that the rudiments of all organic as well as inorganic products and structures must have been evolved in like manner, as they were alike included and contained in the once fused, and therefore once vaporised, or nebulous, mass. In that mass all kinds

Rudiments of all physical things in the primæval mass.

of physical agents, or the elements of them, thermotic, electric, chemical, molecular, gravitational, luminiferous, and by consequence not less all organic and vital forces, must have been included.

Out of it in some way, by equally regular laws in the one case as in the other, must have been evolved all forms of inorganic and equally of organic existence, —whether amorphous masses, crystals, cells, monads, plants, zoophytes, animals, or man,—the *animal* man; the *spiritual* man belonging to *another order* of things, a *spiritual creation.*

Conclusion.

From this brief discussion, which was rendered necessary in order to meet some apparent exceptions to the general view and assertion of *the unity of sciences,* we may now return to the main conclusion, equally valuable in regard to the view it tends to open of the study of the sciences and their relation to each other, as in its bearing on higher inferences which are the crowning pinnacle of scientific truth.

Unity of sciences represents unity of nature.

Sciences in different stages of advance.

All science then is emphatically one: in all its parts and branches, however apparently distinct, or supposed to involve peculiar modes of thought appropriate to each, we find, on close examination, that all such distinctions are but temporary and provisional, and that what appears peculiar is so only because the investigation in different parts of science is in different stages of progress. In one it has arrived at no more than a description and classification of phenomena, or even of the materials whose phenomena we propose to study; in another we have been able to reduce all phenomena to laws of high generality, and those laws to simple principles of force and motion of the most elementary simplicity and the highest generality; and between these extremes there exist all varieties of intermediate stages.

All sciences approach towards showing the unity of nature.

But all sciences approach perfection as they approach to a unity of first principles,— differently applied, indeed, according to the different nature of the material objects contemplated, but in all cases recurring to or tending towards certain high elementary conceptions which are the representatives of the unity of the great archetypal ideas according to which the whole system is arranged. Inductive conceptions, very partially and imperfectly realised and apprehended by human intellect, are the exponents in our minds of these great principles in nature.

The great inference of uniformity is corroborated

not only by the successively more and more comprehensive laws of nature, which science exhibits, but by the very possibility of the existence of such a thing as systematic science: not only by the accumulative proofs existing in nature, but by the marvellous adaptation and harmonising disposition of the human mind for appreciating and discovering them: not only by the occurrence of natural events in invariable order, but also by the possibility of expressing them by laws conveyed in exact terms, and of advancing deductively to the prediction of other phenomena. Thus, even this preliminary condition of all inductive inquiry affords confirmation of the principle of unity of design, connecting the physical with the intellectual world; and this in a still higher degree, as all sciences are seen to tend towards unity.

Our conceptions of natural order the reflection of the reality in the supreme mind.

The actual laws and profound principles which regulate the mechanism of the universe are the originals, the conception and expression of them in the mind of man only the copies. The vast assemblage of physical causes, the great principles, whether of cosmical force or of the minutest molecular affections, as they exist in the heavens or in terrestrial bodies, are the realities: the exposition and demonstration of them in the mind of the philosopher only their images.

All science is but the partial reflexion in the *reason of man*, of the great all-pervading *reason of the universe.* And thus the *unity* of science is the reflexion of the *unity* of nature, and of the *unity* of that supreme reason and intelligence which pervades and rules over nature, and from whence all reason and all science is derived.

§ III.—THE UNIFORMITY OF NATURE.

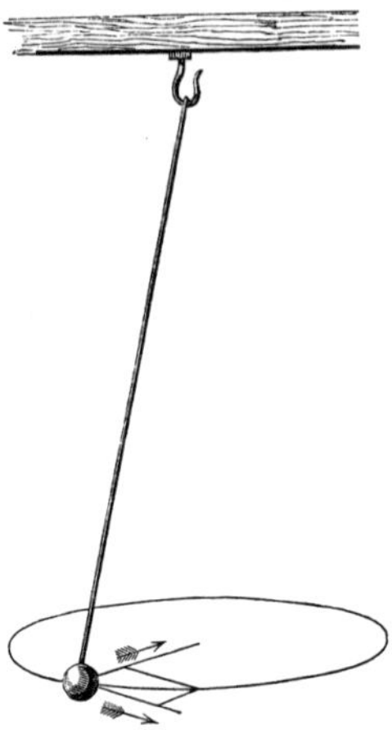

"The harmony of creation is such, that small things constitute a faithful type of greater things."—JEREMIAH HORROCKS, 1637.

Hindrances to science from false analogies.

As real physical analogies form the true ground of inductive speculation, and the power of a right apprehension of them confers that *inductive foresight* which leads to successful discovery, so there are many false views of analogy to be carefully guarded against, involving misconceptions of the relations of physical facts, and leading to mistaken theories and unphilosophical hypotheses, which retard instead of promoting science.

Exceptions and anomalies maintained as real principles.

In the earlier stages of induction, laws are assigned of a limited character, circumscribed by many reservations, and qualified by various exceptions and anomalies, real or apparent, yet which must be at least temporarily and provisionally noted as such. Yet it has sometimes happened that such limited views have been converted into positive and general dogmas, from neglecting the obvious caution of always speaking of them *as provisional.*

Authority of a name.

Thus sometimes, on the one hand, an unduly limited and restricted view, cautiously entertained by a great philosophical leader, may have been caught up by his followers, misunderstood, and invested with a false character and importance; or, on the other hand, crude ideas may be sometimes hastily thrown out by a great master mind, as first temporary or tentative hypotheses, and then come to be treasured up as absolute dicta by their less discerning disciples, and so have acquired the stamp of permanency, to the great and serious hindrance of scientific progress.

Irrationality of dispersion of light.

The once common use of the term "irrationality" of the prismatic spectrum, implied the prevailing prepossession that it might be expected to be "rational," or follow the *same proportion* in all media; whereas, in fact, great difference of ratio prevails. This erroneous first conception long retarded discovery. It, however, shelters itself under the authority of Newton, who assumed it without question, perhaps even as a natural consequence, from the fact, that as spectra given by prisms of all substances are analyses of white light, and can be recompounded into the same white light, the ingredient tints must in all cases be in the same proportion.

Coral reefs.

Nothing could be more marvellous than the hypotheses once universally in vogue as to the formation of *coral reefs*, rising up in so inexplicable a manner from the depths of the ocean; until, by the application of a more correct knowledge of the natural history of the animals, and a simple reference to the common geological phenomena of subsidence, Darwin has divested the whole history of its marvellous character, and restored the equilibrium of inductive uniformity.

When the asteroids were first discovered, an

eminent astronomer maintained that a large planet once moving at their mean distance had *exploded*, of which they were the fragments. Strange as this hypothesis was, it was generally adopted by philosophers, and even calculations were entered into to assign the place at which this marvellous catastrophe took place, and the directions followed by the fragments. But if we simply asked what analogy have we for such an event,—when has a planet ever been known to *burst?* or, indeed, how could such an effect be produced? its unphilosophical nature is sufficiently apparent.

Hypothesis of asteroids from the explosion of a planet.

On the other hand, *condensation* of cosmical matter is an hypothesis which has gained ground from many probable analogies; and the supposition of a ring of such matter, out of which these minute bodies (probably existing in vast numbers) have been condensed, formerly thrown out in a public lecture*, has been sanctioned by the authority of Mr. Adams, in an address from the chair of the Astronomical Society.†

To take another instance: the authority of New-

* Royal Institution, April 7. 1848: see Athenæum.
† Astr. Soc. Notices, 1853, vol. xiii. 143.

Geometrical method.

ton's name and example undoubtedly, for a long series of years after his death, powerfully influenced for the worse the tendencies of the mathematical spirit of England towards an exclusive preference for the geometrical method: or where necessity might compel the use of analytical processes, still an entire devotion to the letter of Newton's fluxional notation restricted their application, and long continued to make the great advances of the continental analysts a sealed book to the English student, and to retard the progress of investigation in this country.

Unity of composition in physiology.

In a similar spirit, without any more disparagement to the great name of Cuvier than to that of Newton, it has been a subject of complaint on the part of a large and increasing school of physiologists, that a too prevalent devotion to the *teleological* methods which he so peculiarly supported, and which derived so undue a preponderance from the authority of his name, has been a great hindrance to the progress of the more extended views opened by the higher principle of "unity of composition" advocated by the school of Geoffry de St. Hilaire, which Cuvier so strenuously opposed, and which the influence of his name was long so potent unduly to repress.

Change of refrangibility.

The prepossession arising from Newton's dictum, "To the same ray ever belongs the same refrangibility," seems long to have operated as a bar to even imagining any theory which involved an opposite idea. And it was accordingly as a sort of paradox that Professor Stokes ventured to announce his important discovery of a change of refrangibility, which affords the key to so wide a range of curious phenomena of light, including and generalising the singular results before obtained by Sir J. Herschel and others of Sir D. Brewster.

Slow progress of generalisation.

The higher and wider extension of analogy and generalisation is not effected at once and at first. The earliest, and sometimes the most highly and justly valued labourers in particular departments and fields of research, as collectors of facts, are not always those best able to perceive the broader connexion of grand principles; and hence are the more apt to cling to such prepossessions as those just alluded to. Even when many classes of facts have been successfully made out, it requires time, and the appearance of some genius of more rare original power, to indicate at once a comprehensive theoretical principle by striking out some general conception, startling, perhaps, from its very simplicity, yet revolutionising the whole science.

Great principles reluctantly adopted.

And even when they are proposed, such more elevated views are not at once appreciated or understood, not merely by the many, but even by the cultivators of science. When Galileo opened the path of all true astronomy by the simple maxim that the same laws of motion which hold good on the surface of the earth apply also throughout the celestial spaces, and when Lyell did the same thing for geology, by maintaining that the analogy of real and existing causes ought to be extended through all the immeasurable periods of past time; — neither was at first admitted without much difficulty and opposition, not so much arising from the mere ignorance of the many, as from the preconceptions of the few.

Dislike of theorising.

Some very eminent men of science have been prone to cherish an intellectual disposition too strongly opposed to all indulgence in hypothesis, and have evinced a very stringent determination to keep to what is regarded as the exactness of demonstrative science, with an especial abhorrence of any thing wearing the appearance of theorising; which they would most carefully shun under the idea of its being metaphysical and visionary, and altogether at variance with the severity of all that is worthy the name of real science, or, at any rate, beyond their province to pursue; which is surely as much a fault in the one extreme as the spirit of fanciful hypothesis is in the other. Yet, men of the most comprehensive minds are the most ready to admit the value of such speculative ideas if well formed.

"Beside positive knowledge," says Humboldt, "stand conjecture and opinion — a philosophical science of nature strives to rise beyond the limited requirements of a bare description of nature. It consists not . . . in the barren accumulation of isolated facts. The curious, the inquiring spirit of man must be suffered to make excursions, . . . still to surmise what cannot be positively known."*

The philosophy of conjecture.

We have already observed that all induction implies a *primary adoption* of a *certain amount of hypothesis;* and the secret of its success in any instance lies mainly in the happy selection of such hypothetical grounds, and not in the mere accumulation of facts. It is by the peculiar capacity for seizing *sound*

* Cosmos, p. 252., 1st trans.

analogies in these first hypotheses that the highest philosophical genius is mainly characterised.

Some persons speak as if all conjecture were alike delusive; but wise and skilful conjectures are very different from hasty and crude guesses; and the comparative probability of several hypotheses, all purely imaginary, admits of many degrees; and to reduce it to something like fixed principles would constitute no unimportant branch of mental science,—the logic of anticipation, the philosophy of the unknown.

Belief from analogy before demonstration.

It was in fact nothing else than the firm persuasion of the truth of great and high principles of philosophical analogy, and the inherent force of sound ideas of probability, which so powerfully influenced those who were the first assertors of the solar system of the planetary world, and even martyrs to its cause, before it had received any absolute proof from the application of mechanical principles; and when the assertion might be called merely conjectural. Yet it was a conjecture of that highest class which is formed by genius in its loftiest moments of inspiration, derived from an enlarged contemplation of the harmony of nature; and, we may add, in like manner it is, that on the other hand the antecedent incredibility of an alleged phenomenon weighs more against it with a sound philosopher, than many assertions in its favour.

Example of Horrocks.

A beautiful example of this kind of anticipation is found in the correspondence of the unfortunate but pre-eminently promising Jeremiah Horrocks; when, after objecting to some theories of Kepler, to account for the planetary motions, he adds, "It appears to me, however, that I have fallen upon the true theory, and that it admits of being illustrated by means of natural movements on the surface of the earth, for *nature everywhere acts according to a uniform plan, and the harmony of creation is such that small things constitute a faithful type of greater things.*" *

It was to illustrate this "true theory," that he devised that beautiful experiment (the most instructive which the lecturer can exhibit even at the present day †) of the freely suspended ball revolving in an ellipse under the combined influence of the central and tangential forces; and in addition showing the motion of the apsides.

Mere conjectural hints at explanation of obscure

* In a letter dated Nov. 23. 1637. See Grant's Hist. of Astron. p. 425.
† See Vignette.

Conjectural explanations only to show that phenomena are not in their nature inexplicable.

phenomena may in many cases be thrown out, which may seem to some fanciful and idle, and may be easily turned into ridicule; when the real object and meaning is only to show that the phenomena in question are not necessarily of such a nature as to be beyond the boundary of *legitimate investigation*, or *possible explanation.* In such instances, to show a *bare possibility* is all that the case requires; the language sometimes employed is perhaps censured as fanciful and evasive, or as a mere disguise for ignorance; but the real purport of such suggestions is overlooked; they are not put forth as pretending to be *complete explanations;* the point that is aimed at is merely to show that the phenomena in question are not necessarily of a kind *outstanding* and *setting at defiance all physical explanation:* now an event cannot be set down as *inexplicable* to our faculties, so long as any possible or imaginable combination of physical conditions can be suggested as capable of furnishing *a plausible explanation* of it.

Example. Nebular theory.

Thus, for example, the nebular hypothesis of the origin of the planetary system was thrown out by Laplace as confessedly a mere conjecture: yet one which was founded on rational probability; and tended to show that the observed peculiarity of the motions of the planets being all in one direction, was *not absolutely inexplicable* on physical principles, and evinced a probability that on this, or some equivalent principle, the origin of those motions might reasonably be expected to find a solution.

Anomaly of satellites of Uranus.

Yet further: to this uniformity in the motions of the system there appears one striking exception in the satellites of Uranus, which are at once *retrograde* and *highly inclined;* as they would be if originally *direct* and then turned over *beyond* the perpendicular. Such a disturbance could not occur from the action of any existing planetary attraction: but, *in the state of nebulosity*, it is far from impossible to conceive some action of the kind among the multitude of conflicting forces then acting. No sound philosopher doubts that the effect was due to *some* regular cause: the nebular hypothesis may serve to suggest that the conception of such a cause is not wholly beyond the limits of physical analogy.

Difficulties in research not to stop it.

The progress of physical discovery may, it is true, be sometimes slow, and the appearance of objections and difficulties so formidable, as to damp the ardour of research, or even to give some colour to the insinuations of those who may be sceptical as to the pretensions of philosophical advance, or entertain jealous or hostile feelings towards such pursuits. But the inquirer, truly imbued with inductive principles, will not despond.

Lesser difficulties not to hinder reception of great principles.

There is one grand maxim of pre-eminent value in philosophic inquiry generally, and which finds a peculiar application under such circumstances as those just referred to, viz., that *having once grasped firmly a great principle, we should be satisfied to leave minor difficulties to await their solution*, assured that in time the progress of discovery will clear them up as certainly as it has now cleared up difficulties once quite as formidable and paradoxes quite as inexplicable. It has been by adherence to this rule that all great philosophical systems have made their way and finally triumphed over error.

Example. Newtonian system and lunar apsides.

The Newtonian theory was beset by palpable contradictions in its results till many years after Newton's death; yet all sound philosophers embraced it. The motion of the apsides of the moon's orbit was, with singular honesty, confessed by Newton to be in fact *nearly twice as great* as calculation from theory made it: and this contradiction remained an outstanding palpable objection, yet without occasioning any misgiving in the mind of sound philosophers as to the general truth of gravitation, till the error was explained, and the calculation rectified by Clairault.*

Up to the present time, *all* the anomalies of the tides are by no means reduced under the dominion of theory: yet no sound philosopher doubts the truth of the principle that they are due to the solar and lunar attractions. **The tides.**

The theory of gravitation, again, was really defective up to the present day: the motions of the planet Uranus, as calculated by theory, were found to be every year becoming more and more discordant with observation, and theory was completely at fault: until the calculation of Adams and Le Verrier showed that the anomalies could be explained by the supposition of the disturbance occasioned by an exterior planet moving in a certain orbit; and, at the time, at a certain point in that orbit; as was directly verified by the observations of Galle and other astronomers. **Motion of Uranus.**

So again, the undulatory theory of light now un- **The wave theory of light.**

---

* Princip. bk. i. sect. ix. prop. 45. cor. 2.

hesitatingly accepted by all mathematicians is still confessedly defective in its application to several phenomena, especially the more extreme cases of prismatic dispersion.

Melloni's anomaly in the solar rays.

The rays of the sun under ordinary circumstances possess a heating power in proportion to their intensity. When analysed, though the heating power differs greatly for different primary rays, and is not proportional to their illuminating power, yet it nowhere exists without rays which may be rendered visible, nor is any visible ray destitute of such power: and for the same ray under the same conditions the heating is proportional to the illuminating intensity.

To this general law one, and one only, outstanding exception occurs in an experiment recorded by Melloni, viz., that, with a certain green glass, the rays transmitted, when concentrated by a lens, are intensely bright, but totally destitute of heat. This is a solitary exception — a breach of all analogy — unsupported by any corroborative experiments: *and as yet unexamined* by any critical experimenter. It is then simply an anomaly *provisionally.*

The anomaly that water is at its greatest density at about 40° F., and below that *expands* with decrease of temperature, has been held by some to be a marvellous and peculiar outstanding fact, setting all theory at defiance.

Maximum density of water.

Yet no truly inductive philosopher for a moment doubts that it is really a part and consequence of some higher law of which the ordinary law of expansion is a part.

Indeed, Berthollet has speculated on the subject, so far, at least, as to maintain that the cause, whatever it be, which produces crystallisation, is in operation in expanding the water before the crystals of ice are actually formed, and which are specifically much lighter than the water.* He even states it as a general law that "the causes which determine the changes of constitution of bodies exercise an action, the effects of which are evident before the changes of constitution have taken place." And this property in water is not altogether an anomaly as compared with what takes place in antimony, iron, and bismuth.

Instances also occur in certain substances in a fluid state which *instantly* solidify on the application of an extraneous body—proving that the particles

* Dr. Pereira's Lectures on Light, p. 168. 2nd ed. 1854.

are already in *a state of strain*, and require only some extraneous agency to bring about that change in their arrangement.

Other suggestions of a theoretical kind have also been made: but, at any rate, we see enough to show that the phenomenon is *not one of necessity outstanding all explanation*, and that it very probably will ere long be brought under the dominion of theory.

Principle of uniformity throughout nature the essence of all induction.

The system of inductive reasoning in its full extent essentially involves the *conviction of the universal and permanent uniformity of nature*. This, as was observed at first, has been emphatically and truly called "*the inductive principle.*" It is this which points to the great archetype of UNITY; to which all our sub sequent conclusions minister increasing confirmation; and from the influence of such a first principle in our inquiries arises all that distinguishes true science from mere empiricism, and an elevated philosophy from the grovelling and mechanical accumulation of mere millions of facts.

Not an intuitive or natural belief.

And we may remark that this idea, in its proper extent, is by no means one of popular acceptance or natural growth. Just so far as the daily experience of every one goes, so far indeed he comes to embrace a certain persuasion of this kind, but merely to this limited extent, that what is going on around him at present, in his own narrow sphere of observation, will go on in like manner in future. The peasant believes that the sun which rose to-day will rise again to-morrow; that the seed put into the ground will be followed in due time by the harvest this year as it was last year, and the like; but has no notion of such inferences in subjects beyond his immediate observation.

Commonly doubted or denied beyond narrow limits.

And it should be observed that each class of persons, in admitting this belief within the limited range of his own experience, though he doubt or deny it in everything beyond, is, in fact, bearing unconscious testimony to its universal truth. Nor, again, is it only among the *most* ignorant that this limitation is put upon the truth. There is a very general propensity to believe that everything beyond common experience, or expressly ascertained laws of nature, is left to the dominion of chance or fate or arbitrary intervention; and even to object to any attempted explanation by physical causes, if conjecturally thrown out for an apparently unaccountable phenomenon.

Acquired only by philosophical study.

The precise doctrine of the *generalisation* of this idea of the uniformity of nature, so far from being obvious, natural, or intuitive, is utterly beyond the attainment of the many. In all the extent of its universality it is characteristic of the philosopher. It is clearly the result of philosophic cultivation and training, and by no means the spontaneous offspring of any primary principle naturally inherent in the mind, as some seem to believe. It is no mere vague persuasion taken up without examination as a common prepossession to which we are always accustomed; on the contrary, all common prejudices and associations are against it. It is pre-eminently *an acquired idea.* It is not attained without deep study and reflection. The best-informed philosopher is the man who most firmly believes it, even in opposition to received notions: its acceptance depends on the extent and profoundness of his inductive studies.

Increasing conviction by more extended research.

Throughout the range which science opens to us we find the several classes and orders of phenomena defined by laws of increasing generality, and thus intimately connected and bound together, so that every part is essential to the coherence and unity of the whole. But when we have arrived at the highest of such generalisations to which science has yet attained, those most comprehensive laws, in the *existing* state of our knowledge, seem diverse, disunited, and not as yet connected by any common higher principle; yet we cannot for a moment suppose this to be anything else than the result of our ignorance; they must each be really subordinate members of some greater group. Future research will undoubtedly connect them together by a common principle, of which at present we can form no more conception than the predecessors of Newton did of universal gravitation, or than he did of electro-magnetism, or geological epochs. Discoveries are being made every day; and the very next important physical discovery will as assuredly effect an union between some two or more classes of phenomena at present not so connected, as the last discovery has done. New *phenomena* are being continually detected. Not more surely does this happen than it is sooner or later followed by the disclosure of more comprehensive *laws*. The progress of discovery is as certain as the extent of nature is unlimited; and the subordination of *species* of phenomena to *genera;* of these genera, again, to *classes* or higher genera, and so on, must be as unbounded as the succession of phenomena.

Idea of chance.

The universality of law and order is the distinguishing conviction of the inductive philosopher; by this, in fact, *science* is elevated into *philosophy.* One main test of its force and extent is the exclusion, in consequence of its admission, of the very *notion of chance,* or of the possibility of any events in the universe really happening *at random.* In fact, the very term chance implies a theory; and if we would examine its meaning, and employ it in a strict sense, we should find that what we really mean can never properly amount to more than a confession of *our ignorance* of the mode or order in which certain events have taken place. If we take any portion of the natural world, or any class of phenomena of which we know least, and which appear most fortuitous, can we correctly say more than that we are ignorant of the laws by which it is regulated? Yet, while in saying that any phenomena appear capricious or fortuitous, we simply admit our ignorance of the laws by which they are governed, no inductive philosopher for a moment doubts that they are regulated by *some* laws.

To take an example: the apparently irregular mode of distribution of the fixed stars through the heavens, might seem, at first sight, to justify the belief that their arrangement and position in the universe might be wholly *fortuitous,* and such probably may be the idea in uninstructed minds, and they may perhaps feel disposed to ridicule the somewhat bold but characteristic idea of Des Cartes*, who says that he meditates an attempt to investigate the *cause* of the position of every fixed star. Yet the very fact that these masses at least have the property of transmitting light to us, and consist of *matter* of some kind, and have been in some instances proved to be subject to the law of gravitation, instantaneously asserts for them a sort of claim of kindred with matter around us and with ourselves, and dispels every shadow of doubt that they are disposed according to *some* physical law, under the influence of some determinate physical forces.

Example. Distribution of fixed stars.

When we come to examine the masterly and profound researches of Struve (disclosed in the "Etudes

Struve's researches.

* Epist. 67.

d'Astronomie Stellaire") we begin to feel more convinced that even in the seemingly capricious distribution of these remote masses through the abysses of space, we obtain a glimpse of *order*, if only from the mere fact that by the consideration of *averages*, some sort of classification is effected; and from that happy combination of arguments brought together from such various sources, which none but an inductive genius of the highest order could have planned, and nothing but consummate mathematical skill could have worked out, conclusions of high generality and profound interest are elicited in a subject, at first sight, seeming to baffle inquiry.

Geological elevations.

Nothing would appear, at first sight, more devoid of all order, or apparently fortuitous, than the directions assumed by those elevations and fractures of strata which diversify the surface of the earth with mountains and valleys, precipices and plains. Yet the accurate observations of geologists, combined with the theoretical indications of dynamical science, have even now begun to throw some light on the probable laws of these seemingly arbitrary manifestations of power, and to connect them with the all-pervading principles of regularity; and though we may not be disposed to assent to the precise theory of Elie de Beaumont, yet it at least gave a right direction to inquiry, and the exact deductions of Mr. Hopkins, place the general dynamical principle of lines of upheaval beyond question; and leave no doubt that a comprehensive mechanical theory will eventually be worked out, and the most monstrous geological "catastrophes" reduced to order and system.

Fate and chance excluded.

Among the ancients we know the several forms of belief in blind fate or chance were not merely popular delusions, but deliberate persuasions, which divided philosophical sects: the advocates of the fixed necessity and eternal destiny of the world, and the supporters of the Epicurean doctrine of the formation of the material universe out of a fortuitous concourse of atoms. But in the age and under the influence of the inductive philosophy, no such dreams can for a moment obtrude themselves. The definite and positive spirit of this system strikes at the root of such vague and unmeaning expressions—the mere disguises of human ignorance. It demands *what* chance and fate are. It appeals to the great principle of *uniformity*, and the regularity of physical causes; and feels warranted in affirming that in all cases, however incapable they may at first appear of reduction to any kind of system, there yet *must be* in reality as perfect, though to us unknown, observance of determinate laws in their production, as in any cases we are most familiar with. Profoundly *adjusted order* is utterly inconsistent with blind destiny, mechanical causes with chance.

No limits to the application of induction.

It is the proper business of inductive science to analyse whatever comes before it. We cannot say that any physical subject proposed is incapable of such analysis, or not a proper subject for it, until it has been tried and found to fail; and even then, the result is not unprofitable; — we know the precise point at which the failure has taken place, and the exact cause of its occurrence. It is a main characteristic of sound philosophy, that it draws the line precisely between the known and the unknown; and teaches us not only why we understand the one part, but why we do not understand the other. Yet the unknown regions on the frontier of science enjoy at least a twilight from its illumination, and are still brightened by the rays of present conjecture, and the hope of future discovery. We can never say that we have arrived at such a boundary as shall place an *impassable* limit to all future advance, provided the attempts at such advance be always made in a strictly inductive spirit. To the truly inductive philosopher, the notion of limit to inquiry is no more real than the mirage which seems to bound the edge of the desert, yet through which the traveller will continue his march to-morrow, as uninterruptedly as to-day over the plain.*

Limits of our present knowledge distinct from those of nature.

When the inductive inquirer finds himself involved in some great apparent difficulty, and among phenomena which no existing resources of science are able to explain, which appear to stand forth as irreducible anomalies, and to baffle all attempts at explanation; however hopeless the problem may seem, he can never really suppose the case to be *in its own nature incapable* of analysis, or that the mass of facts is not really reducible to *some* principles of order, analogy, and causation,—to the dominion of laws as yet indeed unknown, and of causes not as yet conjectured, yet as perfectly regular and strictly harmonious as those which govern the most common daily occurrences,—the fall of a stone, or the ascent of vapour.

* See note at the end of the section.

No *real* interruptions in the order of nature.

A real break in the connection and continuity of physical causes cannot exist in the nature of things. If such breaks often appear, they are due solely to our ignorance. Every advance tends to fill them up; and indeed each physical discovery is nothing else than an extension of the evidence of continuity, a fresh link in the connexion of phenomena into one consistent whole.

Anomalies only apparent.

There is no such thing as any class of phenomena really standing out isolated from all others unconnected by any analogous principle, and truly anomalous in regard to the rest of nature. Yet every class of phenomena has at some time seemed so; but it is an illusion in whatever instance it may now seem to be the case; and one which time will assuredly clear away, as it has already done so many similar or greater illusions.

Anomalies referred to laws as yet unknown.

In all apparent anomalies, the inductive philosopher will fall back on the primary maxim, that it is always *more probable that events of an unaccountable and marvellous character are parts of some great fixed order of causes unknown to us, than that any real interruption occurs.* And further, what may *now* appear the most mysterious, and at present least understood, will yet hereafter be explained by the future extension of discovery.

It may, indeed, be difficult or impossible to apply these considerations *in detail*, and to suggest particular interpretations in subordination to these paramount principles; yet this will not invalidate their general truth; nor need it lead us into extravagant and gratuitous speculations to bring about a precise explanation for which the circumstances do not furnish sufficient data. A truly *rational* inquirer will be content *to let such difficulties await their solution:* and, so far from always seeking such explanations in precise theories, he will admit, on the contrary, that too minute a solicitude to refer every case to KNOWN *causes*, may tend to keep out of sight the broader principle that they may be referable *to some causes as yet* UNKNOWN, but still parts of the same universal order; and may even lead to the disparagement of that principle when, in any instance, such more particular mode of explanation is found to fail.

For example: in the present state of science, of all subjects, that on which we know least is, perhaps, the connexion of our bodily and mental nature, the action of the one on the other, and all the vast Instances.

range of sensations, sympathies, and influences in which those effects are displayed, and of which we have sometimes such extraordinary manifestations in peculiar states of excited cerebral or nervous action, somnambulism, spectral impressions, the phenomena of suspended animation, double consciousness, and the like.* In such cases science has not yet advanced to any generalisations; results only are presented which have not as yet been traced to laws. Yet no inductive inquirer for a moment doubts that these classes of phenomena are all really connected by *some great principles of order.*

If, then, some peculiar manifestations should appear of a more extraordinary character, still less apparently reducible to any *known* principles, it could not be doubted by any philosophic mind that they were in reality harmonious and conspiring parts of some higher series of causes as yet undiscovered.

Anomalies will be cleared up by future discovery.

The most formidable outstanding apparent anomalies will at some future time undoubtedly be found to merge in great and harmonious laws, the connexion will be fully made out, and the

* The reader is referred to "Letters on the Truths contained in Popular Superstitions," by the late Herbert Mayo, M.D. 1849.

claims of order, continuity, and analogy, eventually vindicated.

Inductive philosophy has within itself a prophetical warrant to foresee that a time will come when those things which seem most obscure will become clear. The well-known prediction of such a disclosure in the case of the celestial motions uttered long ago by Seneca *, and fulfilled in Newton, is not less applicable at the present time, and points to equally grand openings in all branches of physical science, which will as assuredly be made at other future epochs of scientific revelation.

No commencement of a new order of things.

When we arrive at any such seeming boundary of present investigation, still this brings us to no *new world* in which a different order of things prevails; it merely points to what will assuredly be a fresh starting point for future research. It is an unwarrantable presumption to assert, that at a mere point of difficulty or obscurity we have reached the boundary of the dominion of physical law, and must suppose all beyond to be arbitrary and inscrutable to

* "Veniet tempus, quo ista quæ nunc latent, in lucem dies extrahat, et longioris ævi diligentia: ad inquisitionem tantorum ætas una non sufficit; veniet tempus quo posteri nostri tam aperta nos nescisse mirentur."—*Nat. Quæst.*, viii. 25.

our faculties. It is the mere refuge and confession of ignorance and indolence to imagine special interruptions, and to abandon reason for mysticism.

Conclusion.

The consideration of the uniformity of nature leads directly to a more precise—a higher—view of the same great conclusion to which we before adverted generally.

Evidence of a supreme mind.

All induction begins and ends in the conception of order, arrangement, and uniformity throughout nature; and this, however inadequately comprehended by our science, is again the evidence of supreme mind, and the universality of order in time and space, the manifestation of the universality and eternity of that supreme mind.

It has been eloquently observed, "Humboldt thought he could show why and how this world, and the universe itself, is a *kosmos,*—a divine whole of life and intellect,—namely, by its all-pervading *eternal laws.* Law is the supreme rule of the universe; and that law is wisdom, is intellect, is reason, whether viewed in the formation of planetary systems, or in the organisation of the worm."*

* Chevalier Bunsen's reply to the President's Address, on delivering the medal to Humboldt, Royal Society Anniversary, 1852.

And in a similar spirit Œrsted has said: "The progress of discovery continually produces fresh evidence that Nature acts according to eternal laws, and that these laws are constituted as the mandates of an infinite perfect reason; so that the friend of Nature lives in a constant rational contemplation of the Omnipresent Divinity."* . . .

"The laws of Nature are the thoughts of Nature; and these are the thoughts of God."†

* Soul in Nature, p. 196.

† Ib. p. 20.

NOTE TO P. 107.

Of the Baconian philosophy, it has been said by a masterly writer, "It is a philosophy which never rests; which has never attained, which is never perfect. Its law is progress. A point which yesterday was invisible is its goal to-day, and will be its starting post to-morrow." (Macaulay's Essay on Lord Bacon, p. 113. small ed.) But while I cannot refrain from citing this brilliant sentence with all the admiration it deserves (and indeed many others in the same essay are not less worthy of admiration), I feel bound to express my dissent from the *exclusively* practical view which the author takes of the objects of inductive science, and must regard it as hardly less than a profanation of the name of Bacon, to associate it with such unmixed and grovelling utilitarianism as he would represent to be its aim.

§ IV.—THE THEORY OF CAUSATION.

1666.

Vel huic philosophandi modo, vel *veriori alicui.*

NEWTON, *Pref. in Princip.*

Desire to seek causes.

AMONG our various intellectual propensities, there is none more powerful or more seductive than the desire to penetrate into the *causes* of things. We perceive events going on or results produced in the natural world; we recognise a number of different powers or agents at work; and to these, under the name of *causes*, or, more strictly, *physical*, or, according to some, *secondary causes*, men are prone in imagination to ascribe a sort of energetic *power*, or a coercive *efficiency*, by virtue of which these physical agents produce, or bring about, certain results: a species of active influence by which matter is imagined to act upon matter, and produce a different state of things, in a way exactly analogous to, if not identical with, that in which a voluntary agent exercises his volition on material objects within his control; and thus there is supposed to exist a relation of a peculiar and intimate, yet hidden and unknown kind, not to be traced by our faculties or further explained, yet the essential condition of all real philosophic investigation; and views more or less similar to these seem to have been very generally entertained among philosophers in former times.

Natural to imagine efficient power.

But when Hume, in his essay on Necessary Connexion, showed that of the existence of this kind of mysterious influence or imaginary power, there neither was, nor could be, *any evidence*;—that in physical events all we could really infer was the mere fact of the *invariable sequence* of the one event called the *effect*, *after* the other called the cause,—a doctrine so

But unphilosophical.

opposed to the favourite mysticism, which delights in investing scientific truth in a veil of abstruseness, and will not condescend to acknowledge any thing intelligible to be true philosophy, was of course not received without much open hostility from some parties; while from others, who felt constrained to acknowledge the strictness of the conclusion, it obtained a reluctant and modified acceptance. It was complained of as a meagre, empty, unsatisfactory doctrine, tending to degrade philosophic speculation to mere matter of fact, and not penetrating below the surface. Thus Lord Kames, though admitting that no connexion of cause and effect is discoverable by *reason*, yet contended that it nevertheless really exists; for we *feel* and acknowledge that every effect implies a cause, and that nothing can begin to exist without a cause of its existence.* That men are prone to *feel* and acknowledge such a notion is perfectly true; but the very question at issue is, do they do so *correctly*, or on any real philosophic ground?

Objections to simple view.

Without here pretending to go into the various

* See Burton's Life of Hume, i. 427.

discussions of the subject which have taken place among subsequent philosophers, and disclaiming all controversy, I will merely remark that at the present day the question is still kept up by advocates of each extreme — the one party contending for the old idea of efficient causation and necessary connexion, and the other adopting the view of Hume, modified by one or two qualifications, yet maintaining the principle of a simple, invariable (or as Mr. Mill terms it, "unconditional") *sequence of events;* and agreeing therein with the French school of positive philosophy, as expounded by M. Comte, in totally rejecting the idea of causation in physical phenomena, in the sense of *efficient* power, as a notion wholly beyond our capacities to define or reason upon, and therefore unphilosophical.

Different opinions at the present day.

My own views of the subject have been expressed in a work published long ago*; but it may be desirable to offer some further explanation of them, after a careful examination of what has been advanced since, whether in support and elucidation of

General view of the case.

* The Connexion of Natural and Divine Truth, London, 1838. See also my Essay on Necessary and Contingent Truth, Oxford, Ashmolean Memoirs, 1849.

the great step made by Hume, or in attempting a retrograde movement and a revival of the exploded, but naturally popular notion, of efficient agency; the personification of matter and mechanical forces, derived from an imaginary analogy between physical action and that of voluntary agents. I conceive that all real philosophical analysis of the case must end in an entire repudiation of such fanciful notions, involving, as they appear to me to do, a confusion of ideas, which I think may be completely avoided by the simple distinction between *physical* causation or the action of *matter* on *matter*, and *moral* causation, or the action of *mind* on *matter*.

Distinction of physical and moral causation.

To take the simplest example: I throw a stone, which brings down a bird; my volition is said to be the *cause* of the stone's flight; the impact of the stone is the *cause* of the bird's fall. The word "cause" is here used in two totally different senses: in the first instance, signifying *moral;* in the latter, *physical* causation. Rejecting altogether the idea of efficient causation, as wholly inapplicable in relation to physical effects, however pleasing to the fancy, I conceive that the true theory of physical causation *includes* the simple idea of an invariable or "uncon-

Physical cause implies sequence in *relation.*

ditional" sequence of facts (meaning sequence in *relation* not necessarily in *time*); yet I contend that there is implied also a *connexion, not* in the *events* in the way of *physical agency,* but in the *reason* and *logical dependence* of the two ideas. The phenomenon or property assigned as the cause or antecedent has undoubtedly a *necessary connexion* with the effect or consequent, when it supplies the *explanation* of it: when the latter is a *consequence in reason* and *theory* from the former,—when, in a word, the cause is a *more general* and *better understood class* or *genus* of phenomena to which we can refer the effect, as a particular *species.*

Necessary connexion *in reason* not in the events.

For example: *friction is the cause of retardation* of motion. There is a mere sequence of two phenomena. Yet there is also a necessary connexion between them, though not in the sense of efficient power; for we conceive the notion of friction, and we then *reason from it,* that retardation will be a *necessary consequence.* But there are many cases where this kind of connexion is less strong and instructive. *Friction is the cause of heat;* but we do not know enough of the nature of friction to be quite certain *why* or *how* it produces heat,

Example: Friction and retardation.

Friction and heat.

though we may *conjecture* it to a certain extent. Here, then, the connexion is not so necessary. Again, *friction* (in certain bodies) *evolves electricity*. Here we know still less of the connexion, and see only a *sequence*. In other words, *physical causation admits of degrees*. But this kind of *connexion* in *reason*, even its highest degree, is totally remote from any analogy with *moral* causation, or the sense of power or effort in a voluntary agent.

Friction and electricity.

Fallacies to be avoided.

"Cause" sometimes used for "occasion."

Cases are sometimes alleged of particular incidental events which are the immediate means, or instruments, or occasions for other events taking place, and are thence called their *causes:* as the opening of floodgates is said to be the *cause* of the flow of water. Yet it is urged gravitation or pressure might, with equal or greater truth, be called the cause; that is, we here use the words cause in a more limited sense. When we speak of physical causes in a philosophical sense, we must recur to the idea not of *mere* sequence of events, but of sequence *in reason*. The pressure of the fluid is doubtless the physical cause of its overflowing: the particular case of floodgates is only an incidental occasion for its action. Some writers, again, fall into, or perplex

Sequence *in time*.

themselves with, what is nothing more than the old fallacy of "post hoc ergo propter hoc;" mistaking a succession *in time* for a succession in *reason* and *relation;* as in objecting that we thus make day the cause of night, and the like.

Order of *time* irrelevant.

In fact, the circumstance of *time is wholly irrelevant* to the idea of cause and effect. We may convince ourselves of this by referring to the numerous instances where the phenomena are *cotemporaneous.*

Cause and effect often coexistent.

Thus the *pressure* and the *density* of elastic fluids are *cotemporaneous* conditions: yet the first is the cause of the second. Evolution of heat with condensation, and absorption of it with expansion, are *coexistent.* Chemical decomposition in the elements of a galvanic battery, and the production of the galvanic current, are *simultaneous.* In these and many similar cases, then, of cause and effect, there is *no sequence at all in time.* The question is as to a *sequence in reason,* and this admits of *many degrees.*

Connexion in reason admits of degrees.

In the last instance of chemical action and galvanism, the effects are *not only simultaneous,* but also *convertible.* Chemical action is the *cause* of galvanism; and galvanism is also the *cause* of chemical action.

This, however, is no contradiction or confusion of ideas: it depends simply on *the relation in which we view the case.* What is meant is, that chemical action, *in the instance of the galvanic battery,* is the cause of the galvanic current; and again, the galvanic current, *in the instance of an experiment performed by that battery,* is the cause of chemical decomposition. We are speaking of different cases.

In some cases cause and effect relative terms.

Thus, in these instances, the use of the terms cause and effect is *relative* to the circumstances and conditions which we are at the time supposing.

Sometimes convertible.

Or again, it is said "magnetism is the cause of electricity, and electricity is the cause of magnetism;" but what is meant is, that in certain experiments, magnetism is so applied as to produce electricity; and in certain *others,* electricity is so applied as to produce magnetism. They are not cause and effect convertibly *in the same sense,* or under the same circumstances. We view them as thus convertible *in different relations.*

Mr. Grove * has considered these cases, and has

* See Correlation of Forces, p. 6.

been led to the conclusion that, "abstract secondary causation does not exist," or, in other words, cause and effect are purely relative terms, *in such cases as those he has considered,* which is exactly what has here been shown.

This is equally true for the other cases which form the subject of Mr. Grove's valuable discussion. The mutual actions of all the imponderable agents, he shows, *are correlative,* or convertible into each other, but no one the essential cause of the other. They are so in different points of view, or on different grounds of relation, as just explained.

In these cases no higher connexion in reason.

But again, as to *the nature of the connexion* between the facts in either case: In the instance of pressure and density of elastic fluid, we perceive *a necessary connexion in reason;* by abstract mathematical reasoning we can infer the one from the other, starting with a definition of an elastic fluid. In the instance of galvanism and chemical action, *we know less* of the connexion, and perhaps cannot show *abstractedly why* one *must* accompany the other.

In the same way in the mutual actions of the other imponderable agents, we cannot reason *abstractedly* to the effects. Where no relation in reason is yet made out, we can only recur to the mere law as experimentally established, and traceable to no higher principle. In such a case, either phenomenon may be cause or effect relatively to the other, as seen under different points of view.

In higher cases no converti-bility.

But in other cases where we have attained a higher and more satisfactory view of a *connexion in reason*, physical causation is more substantially determined. When we can ascend to an abstract principle, and reason conclusively from that principle, that such a result *must* take place as a *consequence* of it, we assign a positive and fixed physical cause in that principle to which we refer; and we cannot reverse the order of relation.

We could not speak of (*e. g.*) gravitation and the tides as cause and effect to each other *convertibly:* or of the connexion of ethereal vibrations and periodical colours as *relative* or *interchangeable.* Whenever we *can* thus mutually convert causes into effects, it only shows the little advance yet made in theoretic generalisation in that particular subject. While, again, in regard to the particular cases of the imponderable agents just considered, it is extremely probable that future discovery will show them to be all merely different modifications of one common principle, and thus easily capable of convertibility in their effects.

Relation of causation to inductive laws.

We thus place the theory of causes in its proper relation to that of inductive laws. In assigning physical causes, we refer a particular phenomenon to a more general,—we refer an *event* to a *law;* and the more strictly we analyse our conceptions the more clearly does it appear that we can never arrive at, or need require, any higher or more intimate connexion than that of successively higher generalisation; by virtue of which to trace a real and satisfactory relation between physical phenomena and the higher abstract principles which combine them together by a "necessary connexion" of *reason*, as parts of a great harmonious whole.

Prejudices opposed to this view.

Yet against this view, it is urged that it is *unsatisfactory;* that the mind still *craves* a more intimate sense of the connexion of events; and that the universal opinion, and common sense of mankind, rejects such cold and dry abstractions, and naturally adopts the more congenial belief in "efficient" causes, and active power in bringing about physical phenomena. This, however, is nothing more than an instance of the general reluctance with which the untaught mind adopts any strict philosophical conclusions: all exact analysis of physical phenomena seems cold and distasteful to the unpractised conceptions, and as men soon dislike what they cannot easily understand, this doubtless is often the origin of the vulgar prejudice and hostility against the higher views of science, and the spirit of abstract philosophy.

General belief no proof.

But even were this persuasion as to efficient causation really *universal*, were it not in fact opposed by as large a section of philosophers as those who uphold it, still, *universal belief* would be no proof of its truth. All mankind, three centuries ago, had a universal belief in the geocentric system. Such general persuasion, if anything, would rather suggest a caution that the popular notion may be a popular delusion. In this, as in another sense, we may say, "argumentum pessimi turba est." And doubtless nothing is more difficult to the unphilosophical mind than to be satisfied with *negation:* to learn the humiliating lesson of its own ignorance.

Idea of efficient cause supposed natural.

Some writers have dwelt upon the idea of causation as arising out of some fundamental principle in the constitution of our minds, and have enlarged on the relation of cause and effect as one under which we are constrained to arrange our perceptions, just as the nature of a machine determines the changes of matter subjected to its action.

But all this is beside the present question, which relates to *what is* the relation in question; and the natural proneness, or necessity, if it be so, is nothing more than a disposition to create in imagination a kind of connexion which does not exist, and to overlook the real and simple relation in which the necessity is simply a necessity of *logical* sequence, applied to a sequence or relation of facts.

The notion of efficient causes is doubtless captivating to the imagination as seeming to let us more intimately into the secrets of Nature. Yet it must be sternly rejected by those philosophers who would adhere strictly to the cautious and positive spirit of the Baconian induction.

But delusive:

In fact, there is an inherent inconsistency in such an appeal to efficient causation. For if this mysterious idea be that which alone supplies a satisfactory insight into the mechanism of the natural world, it must follow, that of the real causes of phenomena we know nothing, even in the cases supposed to be most

and involves inconsistencies.

fully and satisfactorily established: *e. g.* if anywhere surely in the principle of gravitation we must acknowledge a cause which furnishes a complete explanation of the planetary motions; yet the nature of gravity as an efficient cause is confessedly wholly unknown. To the advocates of this view, therefore, the theory of gravitation must be wholly unsatisfactory, and we cannot be said to have attained any real knowledge of the cause of the celestial motions.

Logical connexion of sequence sufficient.

Yet it has been urged by those of this school that the notion of a mere sequence is utterly insufficient, that it is little to say such a phenomenon is produced by virtue of such a *law*,—that a law of action is not action; and the like: nor does the mere reference to a bare *sequence* of events afford any very substantial answer to the objection. The view, however, above explained seems to remove such difficulties. The reference of any class of phenomena to a higher *genus* is really an explanation of its nature: to assign such a governing principle is to show on what the phenomena depends in the connexion of reasoning, which is the only real idea of its necessary relation to a cause.

According to the theory of efficient causes, a species of active power is imagined to reside in natural agents, or to act through them, which constitutes the alleged necessary connection of physical effects with their causes. This is always affirmed to be something of a nature not at all cognisable by our faculties, and dependent on conditions of an occult and mysterious kind.

Efficient causation leads to fanciful physical theories.

Hence it seems to be supposed that anomalous deviations occasionally arise, and the idea of efficient causes is specially favoured by those who are fond of imagining marvellous influences of a kind, distinct from, and even interrupting, the ordinary course of natural events. Such, we must suppose, are the catastrophes and convulsions of nature—failures in creation—random scatterings of matter, and other like notions which are sometimes resorted to as a consolation to the wearied theorist when matter-of-fact inferences seem for a moment to have reached their limit.

Such ideas, however, are not only delusive in themselves, but are radically opposed to the grand truth of the uniformity of nature, the unity of arrangement and design, and by consequence *so far* would tend to impugn the evidence of higher truths.

Opposed to immutability of physical order.

Yet we hear the notion of "efficient causation" in

Confusion of second causes and First Cause.

nature upheld by some as of a peculiarly religious tendency; while (with strange inconsistency) in popular estimation the study of "secondary causes" is accused of being hostile to the belief in a "First Cause." And (from the same confusion of ideas) the denial of efficient causes, and the assertion of a mere sequence of phenomena and laws, is charged with having the same dangerous tendency even in a higher degree.

Charge against the Newtonian system.

Thus, Leibnitz brought against the Newtonian philosophy the strange accusation, "that it deserts mechanical causes, and is built upon miracles, and recurs to occult qualities."*

It seems to have been under the belief of this singular charge that Pope originally wrote the well-known lines which appear in the earlier editions of the "Dunciad,"—

> "Philosophy that reached the heavens before,
> Shrinks to her hidden cause, and is no more;"

which, had the fact been as supposed, would have conveyed as perfectly just a censure in the second line, as it does the characteristic of a *true* philosophy in the first, as *leading to*, not *starting from*, the belief in a Deity.

* See Edleston's Correspondence of Newton, p. 153.

Whereas, when undeceived as to the fact, the lines which he substituted in the later editions,—

> "Philosophy that leaned on Heaven before,
> Shrinks to her second cause, and is no more,"

embody the whole vulgar misconception and confusion of ideas respecting First and Second Causes, while they are, in any sense, as wholly inapplicable to the *Newtonian* philosophy, as the former.

Difficulties removed.

A recurrence to (what is at least) the simple and intelligible view above expounded, would remove altogether the whole mass of difficulty, confusion, and objection, in which we are thus entangled, and which is involved in the notion, so commonly alleged, of an eternal succession of secondary causes, excluding the idea of a First Cause, and the like. If we say that every event must have a cause, it means that every species of phenomenon belongs to a class more comprehensive; that class to a still larger, and so on. The "*summum genus*" of all (if any induction could reach it) would be nothing else than an ultimate *physical* principle of the whole universe; but would still be so far from trenching upon the idea of a supreme *moral* Cause, as to be, on the contrary, the very highest and crowning proof of the influence of *mind*, in the evidence it would give of the ultimate principle of *universal order*.

Conclusion. The connection and subordination of inductive laws and generalisations is what we carefully distinguish as *physical causation*. But material *unity*, *system*, and *order*, are the indications of *mind*; and the connected series of *physical causation* is the manifestation of *moral causation*.

Thus, the truly inductive philosopher recognises *presiding Mind*, the supreme *moral* Cause of all things, everywhere revealed by the same outward manifestations of universal order and harmony; everywhere indicated by the same external attributes, symmetry, uniformity, continuity; and attended by the same ministering agents, invariable laws, and physical causes.

---

* Hume's view of causation was censured by some of his opponents as leaving the connexion of all events so loose as to open the door to the supposition of causes sometimes failing to produce their effects or effects occurring without causes, or of all things being abandoned to chance or destiny.

Though, as has been well observed by his biographer, Mr. Burton, such objections are of a vulgar class, and not such as a philosopher would entertain, yet it may be worth noticing how completely the possibility of falling into such absurd misconceptions is avoided by the view taken above.—Life of Hume, i. 81.

NOTE TO PAGE 116.

It should be observed that the opinion quoted of Lord Kames, besides the objection noticed in the text, involves also an instance of the entire confusion of the idea of physical and moral causation here dwelt upon.

§ V.—FINAL CAUSES, AND NATURAL THEOLOGY.

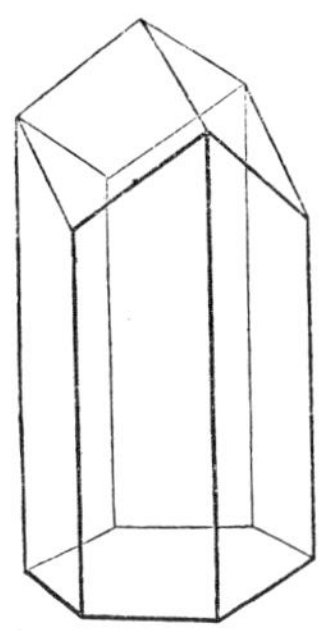

κατά τινα γεωμετρίκην μηχανῶνται προνοίαν.

PAPPUS.

THE theory of causation has been much mixed up with the argument evincing *design* in the arrangements of nature; and under the name of "Final Causes" that argument has been involved in no small confusion of ideas: and notwithstanding much which has been urged on the other side, still, along with Dugald Stewart and others, I cannot but agree in thinking that the term "Final Cause" is most unhappily chosen to express the true meaning, and

Final causes.

The term ill chosen.

has tended to convey an idea not only too limited, but altogether confused and misleading.

Nor can I doubt that much of the obloquy which has been cast on natural theology generally by writers of a sceptical school, has arisen out of the narrow views thus implied, especially when the argument has been almost wholly restricted to physiology, and the very idea of intention represented as the essential characteristic of *organisation,* and this branch of science imagined to involve principles different from those prevailing in other branches; all which might therefore naturally be imagined barren of such application.

Objection from too narrow a view of final causes.

It is, I conceive, solely as being understood in this narrow sense, that "final causes" are so vehemently assailed by Comte and writers of his school; and it cannot be denied that among the advocates of natural theology there exists too common a disposition to *narrow* and *restrict* the application of the argument by *confining* the proofs of design to those instances *of adaptation of means to a perceptible end* of which we doubtless find such abundant instances throughout organised nature, instead of taking a more expanded view.

The instances in which we can trace a *use* and a *purpose* in nature, striking as they are, after all constitute but a very small and subordinate portion of the vast scheme of universal order and harmony of design which pervades and connects the whole. Throughout the immensely greater part of nature we can trace *symmetry* and *arrangement,* but not the *end for which* the adjustment is made. But this is in no way a less powerful proof of design and intelligence than the former. The most exact and recondite adaptation of means to accomplish an obvious end is *no more peculiarly* an evidence of design, than the universal arrangement according to determinate laws which pervades the depths of cosmical space,—where we are least able to trace any end. Symmetry and beauty are *results of mind* of at least as high an order as mechanical efficiency. A mere numerical relation invariably preserved, but no further connected with any imaginable purpose, or a systematic arrangement of useless parts or abortive organs on a regular plan, are just as forcible indications of intelligence, as any results of immediate practical utility.

Means adapted to an end not the highest idea of design.

Symmetry a proof of design.

That the one class of results are more immediately

Proofs of design equally in symmetry and in an end answered.

striking and obvious to minds of every capacity with little study or inquiry is true; and that the appreciation of the claims of symmetry even in unorganised structures, the harmony of profoundly adjusted laws, and the conspiring tendency of all these laws towards a grand pervading unity throughout the physical world, is not so easily caught, and is perhaps fully appreciable only by the cultivated and philosophic mind, is equally true. But this is a difference merely in degree, and as applicable to different classes of minds. In a philosophic analysis of our convictions there can be no real difference in kind between the two classes of conclusions.

In a strictly philosophical point of view the inference that everything "has a use" may certainly be regarded as a generalisation which carries with it a high degree of probability. We find that *many* things have a manifest use; but then we find innumerable others for which we can discover no use, and by which no visible end or purpose is answered, but it is not an unfair extension of the inference that, in these cases, some unknown end is answered; that, in fact, everything in nature is adapted to other things and to the whole; though in

In most cases no purpose traceable.

by far the greater part of nature we fail to perceive *what* the particular relation or dependence may be. It may suffice to convince us of this, if we merely ask for what *purpose* is life itself conferred? or, to what *end* does the material universe altogether exist?

In some, purpose apparently defeated.

Again: the usual argument for design in organised structures is, that the various adjustments point to the designed end of life and enjoyment to which they are subservient; but it is an obvious objection that these ends in numberless cases are *not attained;* there is malformation and suffering, disorganisation and disease; and, finally, the whole design is always defeated and put an end to by death. It is hence manifest that to take a satisfactory view of the case, we must not rely on the *mere* consideration of an *end answered,* but must recur to a higher principle—that of symmetry, order, unity of plan, and composition of organised frames: and this too, as only one branch of the yet wider scheme of universal order.

But in certain cases an end answered when not obvious.

It is, however, fairly to be admitted, that many instances occur where we should least expect it of utility in natural arrangements. Thus it is argued by some that an *apparent waste* is not necessarily a *real* one; for the sustenance of one single species or individual, a numberless combination of conditions must co-exist; if one condition were altered, all the rest would fail in their co-operation, and the individual, or even the whole species, would perish. Thus all, even the most apparently remote, arrangements of things which seem to have no relation to animal life are yet essential to it; and thus the barren desert and the void ocean are not *wasted*, but essential parts in the economy of the minutest forms of animal life in the most distant hemisphere of our globe.

In the same point of view Œrsted has beautifully observed, "There is no inactive void in the remote distances between the planets. The space is filled by ether, and is penetrated by the attractive forces by which the whole universe is held together. The ether itself is an ocean whose waves form light, that great connecting link which conveys messages from globe to globe and from system to system."*

Yet the least consideration shows that we must not press such arguments beyond their due limits;

* Soul in Nature, p. 55.

and still less make them the exclusive view of the subject.

Views of the ancients on final causes.

Though the ancients* reasoned justly and admirably on final causes in certain familiar instances, and in the limited sense of the adaptation of means to a known end, yet the state of their physical philosophy absolutely prohibited wider views of unity and order. Under a system which could not go beyond the assignment of each class of phenomena to some peculiar unknown efficient cause, unconnected with others, no such generalisation as *unity of design* could have been legitimately attained.

Final causes misplaced in philosophy.

The remark of Bacon† that final causes are not in themselves to be rejected, but have been *wrongly placed* in philosophy, is one of more value than seems generally understood. It may be very true that sometimes hints towards inductive investigation have been obtained from the consideration of the ends to be answered by certain observed conditions. But it is in general a more safe and philosophical rule, that we may in all cases argue *from* physical

* We cannot have a more striking instance than in the well-known and justly admired passage—Xenophon, Memorabilia, i. 4.

† De Augmentis, lib. iii. c. 4.

inductions to final causes, but not *from* final causes to physical inductions.

Improved views now required.

The old and limited view of final causes will not meet the increasing demands of scientific enlightenment; it will not suffice now to argue solely on the adaptation of means to a known purpose, or a practical design evinced, and an obvious end answered. If we cannot discard the term, we must enlarge its meaning. We may speak of "design" with reference solely to "order" and "arrangement," without looking to the idea of *practical* utility. Such modes of expression are far preferable, as not leading the mind to any undue expectation of what it will not realise.

Extension of argument for design to unity of composition in physiology.

Thus in reference to physiology, the higher argument acquires an expansion in proportion to the progress of the science. We obtain more enlarged ideas of *design* as we advance from the more confined views of the older schools towards the wider principle of *symmetry and unity of composition.* So that "final causes" *properly understood*, so far from receding (as some pretend) before the advance of modern science in the wider and more philosophic sense, eminently derive increasing evidence from its progress. The study of the higher principle of *symmetry* and unity of composition can in no way prejudice that of *adaptation*; the latter being but a part of the same great argument. Nor is it just to accuse those of the modern school who are engaged, as their special and legitimate object, in investigating the former, of undervaluing the latter.

The cells of the honey-comb.

The celebrated case of the cells of bees deserves more particular consideration, inasmuch as it offers an instance in which the proof of *mind* is independent of the idea of mere utility. It is scarcely necessary to observe that the supposition adopted by some of a mere pressure upon a cylindrical cell producing the hexagonal form is wholly insufficient: the main point to be accounted for is the highly artificial mode of *termination* of the cell by three rhombs* inclined at the precise angle (70° 31′) which calculation requires for the minimum surface, which is also the acute angle of the rhomb. The argument points to a highly intellectual operation either performed by the bee, or implied in the arrangement of its organs, so as mechanically to effect it. On either alternative the proof of *mind* is independent of the consider-

* See Vignette at the beginning.

ation of a useful *end answered:* it depends on the conception and solution of what is to our intellects an abstract mathematical problem, by no means of an elementary or evident nature; and which is equally remarkable whether any *purpose* were *fulfilled* by its application, or not.

Argument from astronomy.

Paley expressly held that the mechanism of the heavens was a branch of science the least susceptible of this kind of application: according to the principle here advocated, it forms the highest and most satisfactory.

Stability of the planetary system.

But a more special argument has been raised on the ground that the planetary perturbations have been shown so to compensate each other, that no permanent derangement can arise; and Laplace pointed out that this stability of the planetary system is the necessary consequence of certain conditions, *not themselves necessary;* viz. the smallness of the inclinations and eccentricities, the motions all in the same direction, the comparatively vast mass of the sun, and the incommensurability of the periods.

Professor Playfair* justly enlarges on this as an

* Playfair's Works, iv. 294. 318.

argument for *design;* but if the conditions thus assigned were necessary (i. e. *necessary consequences* of each other or of something else), he thinks we could not infer *design.* They, however, are not *necessary:* each *might be* otherwise, the rest remaining. Their existence then, he argues, not arising from *necessity,* nor from *mechanical causes,* nor from *chance,* must be from *design* and intelligence.

But I would ask, Suppose *they were necessary* consequences of each other or of some higher principle, or did arise from mechanical causes, would not that higher principle, or those causes, so arranged as to produce them, be an *equal* proof of *design,* or even a *higher?* So singularly deep-seated is the prejudice, that design can only be inferred when we cease to trace *laws,* or when conditions appear *arbitrary.*

Mechanical necessity.

The idea of "mechanical necessity" (derived probably from the school philosophy) as something distinct from the result of *systematic plan* in the order of the universe, has long continued to haunt the ideas of writers on the subject, and to be the source of many cavils.

Thus in past times the Newtonian discoveries were accused by many of having an irreligious tendency in reducing everything to "mechanical necessity." And even so enlightened an advocate as Cotes,* instead of showing the fallacy involved in that very term, replies by contrasting "necessity" with "design," when it might have been pointed out that such *necessity of reason is the highest proof of design.* Other philosophers we find sometimes questioning whether certain results may have been brought about by the direct interposition of "the First Cause," or by some unknown "secondary cause," as if the two were *opposed* to each other; or, as if science could have any evidence of the first except through the channel of the second.

Uniformity of natural causes.

From the inductive philosophy we derive our belief in the harmony, order, and uniformity of natural causes, perpetually maintained in a universally connected chain of dependence. And hence it is, that we arrive at those sublime ideas of a presiding Intelligence of which *law* and *uniformity*, universal mechanism once for all adjusted, are the proper *external manifestations.*

---

* "Naturæ leges . . . in quibus multa sane sapientissimi *consilii*, nulla *necessitatis*, apparent vestigia."—Pref. to 2d edition of Principia, (p. xxix.).

To the truly inductive philosopher, *fate* and *chance*, *necessity* and *accident*, are words without meaning. To him, the world is made up of recondite combinations of physical laws, and the existence and maintenance of those laws are the very indication of a Supreme Mind. But chance is irreconcilable with laws, fate with mind, regulated and fixed order with blind destiny, fortuitous accident, or arbitrary interruption.

Natural theology strengthened by the chain of causes.

All rational natural theology advances by tracing the immediate mechanical steps and particular processes in detail, and the physical causes in which the influences of the Great Moral Cause or Supreme Mind are manifested. The greater the number and extent of such secondary steps and intermediate processes through which we can trace it, the greater the complexity and wider the ramifications of the chain of causes, the more powerful and convincing the instruction they convey as to the existence and operation of the Divine wisdom and power.

Mistaken ideas.

Yet it is a common mode of illustration to speak of the *chain* of secondary causes reaching up to the First Cause. Or, again, fears are entertained of tracing secondary causes too far, so as to entrench on the supremacy of the First Cause. But this is an erroneous analogy: the maker or designer of a chain is no more at one end of it than at the other. The length of the chain in no way alters our conviction of its skilful structure, except to enhance it. If the number of links were truly *infinite*, so much the more infinite the skill of its framer.

Mr. F. Newman* observes, I think most truly, that the *common* arguments from what are called "secondary causes" to the "First Cause" are unsatisfactory: and I would trace this to the confused sense in which those terms are commonly used, as already explained; and which, I think, might be entirely removed by attention to the distinctions above laid down. While, on the other hand, I fully acknowledge that those arguments, when correctly understood, lead only to a *very limited* conclusion; and one which falls infinitely *short* of those high moral and spiritual intuitions on which Mr. F. Newman grounds his religious system, yet in no way discredits them.

Again, by some well-meaning but confused rea-

* Soul, p. 35.

soners, the argument is often put in reverse order; and so stated as to appear as if the *assumption* of a Supreme Mind or an "efficient and intelligent Cause" were really the *basis* of our belief in the uniformity of nature, instead of the *conclusion from it.* Yet if this were so, what would it be but to render the whole proof of a Deity, *an argument in a circle?* So, in like manner, some would set out by insisting on the idea of "a purpose answered" and "an intention" as an *essential antecedent* part of our conception of an organised being: and then, from the study of organised beings, would *deduce* the *conclusion* of design and intention!

Confusion in the order of argument.

Thus Coleridge observes, "Assume the existence of God, and then the harmony and fitness of the physical creation may be shown to correspond with, and support, such an assumption: but to set about proving the existence of a God by such means, is a mere circle,—a delusion!"* Now I would observe that for the theological idea of God, the natural argument is no doubt insufficient, but still it is no argument in a circle; it is strictly logical as

* Table Talk, p. 307.

far as it goes, though that is but to a very limited extent.

Again, the same author asks, "How did the Atheist get his idea of that God whom he denies?" * The answer is unhappily obvious, that he usually takes it up from the narrow and unworthy representations of dogmatic systems or puerile recollections, instead of the ideas suggested by a sound philosophy.

Mistakes arising from ideas of causation.

Among some writers of an eminently religious spirit at the present day, we cannot but notice the unhappy influence of that confusion of ideas on the subject of *causation*, as well the want of due appreciation of the grounds and nature of physical philosophy in reference to the inferences of natural theology, which it has been the object of the foregoing remarks to obviate.

Thus Sterling † observes, "Physical results prove nothing but a physical cause." Again, "It is thoughtless to say that, because all things we know have each their cause, therefore the whole must have one cause." . . . . . "Every phenomenon *within nature* has a cause; but this does not entitle

* Table Talk, p. 307.

† Essays, ii. 121, 122.

us to go beyond and look at nature from without, and say this, too, must have a cause."

The real ground he maintains is very different, and is suggested by the question, "Why is the view of the universe so weary, fearful and unsatisfactory? The sense that we need a God is an infallible indication that there is one" . . . . . &c.

After what has been before observed, it is hardly necessary to point out, in detail, how completely all this perplexity would have been removed by better views of physical philosophy, the want of which, so serious in discussing a subject essentially founded on physical truth, the author's literary attainments, poetic imagination, and high devotional feeling, were incompetent to supply.

The very idea of "looking at nature from without," evinces this perplexity; and the unhappy view of the universe as "weary, fearful, and unsatisfactory," would be banished at once by the juster contemplations of an enlarged inductive philosophy, investing the whole with the cheering light of universal beauty, order, and harmony.

If, indeed, the author meant simply to transfer the belief in a Deity altogether from the domain of *reason*, to place it in that of *feeling;* to ground it on the sole consciousness of internal emotion, or the intuitive impressions of individual experience, this would be a view to which the philosophical argument *offers no disparagement*, though it does not reach up to it.

Erroneous notions of the limits of nature.

Some, however, would assert, that after all physical explanations, there remains the same ultimate incomprehensibility in natural causes; and that even in nature we find ourselves surrounded by wonders and miracles: ideas which only evince a total absence of distinct philosophical thought, and confound the limits of *nature* with the limits of our PRESENT *knowledge* of it — *unexplained* phenomena with *violations* of physical order.

They are fond of speaking of the *limits of nature*, of a region of inscrutable *mystery* by which the frontiers of science are on all sides surrounded, impenetrable to our faculties, and forbidding advance.

No hindrance to advance.

If by mystery they mean something into which we neither can, nor ought, to inquire, then in accordance with what was before observed,* in science there

* See above, p. 106.

are no mysteries, no inductive inquiry can ever bring us to such a termination.

If we limit the term "nature" to that portion of the universal fabric whose laws and mechanical causes are more or less perfectly known to us, the distinction is then merely incidental and fluctuating: it is purely *relative* to ourselves and the temporary extent of our knowledge, and presents no really essential difference, being dependent merely on the extent to which the boundaries of knowledge are pushed forward at any particular epoch.

Nature means universal order.

But "nature," in its wider sense, implies a whole, all of whose parts are united by a community of character; and no one portion of it, whether known or unknown to us, can really be beyond those ordinances of recondite arrangement, a small portion of which is manifested to us.

To assert an arbitrary condition of things whenever our inductions fail, is to place such cases beyond the boundaries of design: to suppose a region of mysterious confusion beyond all law and order, is to discredit the universal influence of Supreme Mind.

Limited notion of "nature" as opposed to "Deity."

In accordance with the narrow and unworthy notions formerly prevalent on these subjects (perhaps inseparable from an earlier stage of science), it would seem to have been held, that the appearances of the physical world, so far as they were reducible to regular laws, were to be regarded as what was termed "Nature." When we reached the boundary of the province thus subject to reason (as we soon must do in any direction), and when phenomena seemed in any instance not so reducible to laws, then we arrived at the limits of "nature," and were reluctantly compelled to resort to a Deity, a *θεος απο μηχανης*,—a Supreme Being admitted on compulsion, when the order of things could no further be traced without Him. Then, and not till then, we might exclaim with the poet, "Ergo perfugium."*

Erroneous argument from planetary perturbations.

Thus, to take an instance, minds incapable of appreciating Newton's own sublime inference from the uniformity and order of the system which he had so marvellously and happily disclosed, have dwelt upon his single expression (when in ignorance of the extent and fertility of his own principle in leading to the great law of stability), that at length the increase of planetary perturbations would require a special inter-

* Lucret. v. 1185.

vention for restoring the equilibrium. This was applauded as the only satisfactory acknowledgment of a Supreme Power. We merely ask, *If this be the true argument, what now becomes of the conclusion?*

Certain phenomena in geology.

Just in the same way we hear (for example) religious writers at the present day arguing on certain obscure and unexplained phenomena of geology. They find indications of what may seem *apparently* abrupt changes in the orders of organised beings in past times; and because no established law or physical theory will immediately apply, to assign an adequate cause (supposing the fact to be so), these changes are triumphantly adduced as the special footsteps of the Creator (as if the whole of geology presented anything else); so that when future and enlarged discovery shall disclose the connection and explanation of these appearances by regular laws, their argument for a Deity will fall to the ground!

In physics.

According to this mode of representation, "nature" was the *rule*, "Deity" the *exception.* The belief in nature was the doctrine of reason and knowledge; the acknowledgment of a God was only the confession of ignorance. So long as we could trace physical laws, nature was our acknowledged and legitimate guide; when we could attain nothing better, *we were to rest satisfied with a God!* Even learned writers on natural theology have thought it pious to argue in this way. To take a single example: The apparent anomaly that water arrives at its maximum density before freezing, occasions its *freezing first at the surface,* and other results connected with important points in the economy of the globe and the good of its inhabitants: and this argument for design is sometimes represented as if it acquired a *peculiar force* from the circumstance of the fact being an *anomaly,* and inexplicable by our theories. And *on this ground* it is particularly held up to popular acceptance as an instance of special intervention, for the benefit of man, traceable to no physical cause. But when the apparent exception shall come to be reduced to its proper place as a part of some more comprehensive law (as it assuredly will),* all the peculiarity and mystery of the case will be at an end, *and with it will fall the* theological argument and the popular faith, propped up on so false a support.

Influence of prejudices.

Yet in spite of the better knowledge which ought to

* See above, p. 96.

prevail, we often hear, for example, any sudden and marvellous infliction of disease or famine, pestilence or blight, which (it is added with a sort of triumph) "baffle the boasted powers of science to explain," held forth as signal instances of direct interposition.

To resort to such representations, however it may serve a temporary purpose, or exert an influence on the multitude, is the resource of ignorance, the encouragement of superstition, and eventually the unfailing parent of a sceptical and irreligious re-action; and if the faith of the many be propped up by such false supports, it must fail altogether as soon as increasing knowledge clears them away.*

Interruptions of order interruptions in the evidence of design.

To speak of apparent anomalies and interruptions as *special* indications of the Deity, is altogether a mistake. In truth, so far as the *anomalous* character of any phenomenon can affect the inference of presiding Intelligence at all, it would rather tend to *diminish* and detract from that evidence. But, on

* In relation to this subject, I cannot refrain from quoting a single sentence from a discourse of a very opposite tendency to such as I have just alluded to, and of a kind which it is to be wished were more common: —

"God punishes us . . . . not by His caprice, but by His laws. He does not break His laws to harm us; the laws themselves harm us when we break them and get in their way."

"Who Causes Pestilence?" Four Sermons by the Rev. C. Kingsley, Rector of Eversley, p. 14. London, 1854.

the other hand, precisely in proportion as the apparent exception might be explained, and made to vindicate its position in a more comprehensive system of order, so would the evidence be increased and elevated.

In the present state of knowledge, law and order, physical causation and uniformity of action, are the elevated manifestations of Divinity, creation and providence. Interruptions of such order (if for a moment they could be admitted as such) could only produce a sort of temporary concealment of such manifestations, and involve the beautiful light shed over the natural world in a passing cloud. We do not indeed *doubt* that the sun exists behind the cloud, but we certainly do not see it; still less can we call the obscuration a special *proof* of its presence. The main point in the system of order and law is its absolute *universality*. Exceptions, if real, must *pro tanto* imply a deficiency in the chain of connection, and might, to a sceptical disposition, offer a ground of doubt.

But so overwhelming is the mass and body of proof, that no philosophic mind would allow such exceptions for a moment to weigh against it; they would be as dust in the balance. A supreme moral cause manifested through law, order, and physical causes, is the confession of science: conflicting operations, arbitrary interruptions, abrupt discontinuities, are the idols of ignorance, and if they really prevailed, would so far be to the philosopher only the exponents of chaos and atheism; the obscuration (as far as it extends) of the sensible manifestation of the Supreme Intelligence.*

False views from ignorance of inductive principle.

That idea was hidden in former ages, ignorant of the true inductive philosophy,—in proportion

* The question of apparent interruptions has been much discussed by Œrsted, in his work already referred to. (See Soul in Nature, pp. 59. 173. 178.) Some, he says, imagine accidental causes of derangement in nature which may render arbitrary intervention necessary for their readjustment, as was once supposed with regard to the planetary perturbations; but, as knowledge has progressed, it has been more clearly seen that the error lies in supposing such deranging causes accidental; they are all results of the same general laws, modified by particular conditions. He puts the parallel of human contrivances, in which, in proportion to the skill and intelligence employed, such derangements are foreseen and provided for, as *e. g.* the compensation for expansion in chronometers; and even in moral agency, as in state institutions providing remedies for the lawlessness of criminals. (173.) Infinitely more then, he argues, must we expect such provisions in *Infinite Wisdom*, which is sufficient to guide everything without requiring alteration. (178.) Apparently inexplicable events are so only to our general ignorance. (59.) Real interruptions would suppose deficiency of reason in God.

On this point the author has forcibly remarked, "When our opponents triumphantly bring forward inexplicable events, we can reply to them, In common with us you cannot understand these events, but you fancy you understand them; you believe that you are initiated into God's decrees, and speak accordingly: we know that we do not understand them, and openly declare it. They may perhaps assert that they are guided by religion; that they judge by the will of God revealed to them by religion: but only let them show us a single instance of an event where it can be applied without the addition of some of their own wisdom."—Ib. 178.

"We often hear it said that some things would be inexplicable if we did not believe in higher arbitrary arrangements; but that anything is inexplicable without a certain presupposition is generally a very weak proof of its being really so."—Ib. 180.

as men speculated on false principles, and imagined (as was very natural) a plurality of supernatural powers, and unconnected or opposing influences, whose conflicting operations were evinced in the mysterious irregularities and capricious course of natural phenomena. In earlier stages, even of inductive inquiry, though its higher principles were in some measure recognised, it was yet supposed that limits existed to their dominion. And even at the present day we cannot say that such a notion has been generally or absolutely exploded.

Narrow views of nature to be enlarged.

Among philosophers, though the idea of a limitation has been slowly dispelled, yet still many are unable fully to embrace the universality of law and order, or to attach to it the high importance which it rightly claims.

In a word, according to the narrow, but prevalent idea, the great universe is nothing but an immensity of arbitrary and inscrutable darkness and mystery, to which the philosopher's inductive world — a limited portion of matter, chained down by mechanical causes — forms an insignificant exception.

But the more worthy conception looks to a boundless cosmos, a universe of order, a grand scheme of eternal laws adjusted by Supreme Reason, of which our limited inductive knowledge opens a partial glimpse, yet calculated to convey a faint impression of that immensity of Intelligence which pervades, animates, and rules not only our sphere, but all beyond it.

Improvements in science advance natural theology.

Improved views, increased and accumulating evidence of the harmony pervading the material world, are attained in proportion to the advance of sound inductive science. The more close adherence to the spirit of philosophical analogy leads to a more commanding sense of the uniformity of nature, and the true idea of causation. As the generalisations of physical science become more comprehensive, we acquire juster notions of the stupendous aggregate of physical causes, of the inconceivable vastness and complexity of that universal mechanism, some small portion of which we are enabled to understand; and whose recondite and perfect adjustment, however imperfectly perceived, is the true ground and evidence of our conceptions, partial and limited as they must be, of the Infinite Source of all things.

Higher views of science important,

It is not a mere desultory and fragmentary knowledge of detached facts and portions of science, to however great an extent it be carried, which can suffice to lead us to a correct perception of those truths. It can only be by a thorough insight into the interior principles of the inductive philosophy, and an imbibing of its real spirit, that we can attain adequate perception and sense of the real unity of nature which forms the basis and substance of those more sublime inferences.

Leading to higher views of natural theology.

In the confined and literal notions, often ignorantly entertained, of the sciences of observation, our conclusions might be supposed restricted to the field of mere sensible experience; and in this sense we should fall short of any worthy apprehension of the Supreme Intelligence. But the truly inductive philosopher extends his contemplation to intellectual conceptions of a higher class, pointing to order and uniformity as constant and universal as the extent of nature itself in space and in time; and in the same proportion he recognises harmony and arrangement invested with the attributes of universality and eternity, and thus derives his loftier ideas of the Divine perfections.

The real nature and bearing of the evidence of natural theology as founded on *universal order*, has in fact come to be better understood only in an age of advanced philosophic cultivation: it tends to become continually more perfect with increasing knowledge; and its full force is hardly yet commonly apprehended even among men of science.

In ignorant ages all phenomena viewed as supernatural.

The stupendous phenomena of nature are indeed the manifestations of the Supreme Power as well to an ignorant, as to a cultivated age and people, though the impression is produced in a very different way, and excites a very different tone of feeling. In an age when these phenomena have not been reduced to laws, or traced to causes, they are all ascribed to arbitrary influences. When the uniformity of nature was unknown, the violations of it offered no contradiction. When everything was supernatural, no discrimination of evidence was possible. In those ages all phenomena, whether in the inorganic world, or in varied influences on the human constitution, were unavoidably regarded as direct interventions and acts of the Deity, and were truly described as such in the writings of those periods.

Natural theology of ruder ages.

In the ruder stages of man's progress, religious impressions are more peculiarly produced through the medium of the feelings of awe and astonishment, which are called forth by the occurrence of the more rare, extraordinary, and fearful phenomena, prodigies and marvels. These joined to the more tangible influence of events on their own fortunes and enjoyments, believed to be retributive judgments and providential deliverances, when favourable to themselves and destructive to their enemies, are the only appeal to which men were then accessible.

More enlightened views.

But with the enlightenment of physical discovery, a more definite natural theology presents us with conclusions which, though resting on an unassailable basis, are restricted in their character and extent.

Limits of natural theology.

Natural theology, as based in physical science, confessedly leads us only to a very limited conception of the Divine perfections; it traces beneficent arrangements yet mixed with a large proportion of evil; it recognises omnipotence in the constitution of the immense connected machinery of the universe, and the perpetual maintenance of its determinate laws, rather than in their interruption. At the very utmost it points to providential government in the preservation of an unbroken system of pre-ordained causes for the general good, rather than it suspension for the benefit of individual parts, and influence on mind rather than disarrangement of matter.

Spiritual views from other sources.

If the human mind or human desires require fuller manifestations, or aspire to a higher sense of the Divinity, it must be from *other* and *more spiritual sources* that such wants can be satisfied,—a philosophic natural theology while it cannot furnish such satisfaction, yet at least puts no hindrance in the way of its attainment from other and more appropriate teaching.

Argument from design to a designer.

But the great argument which we have been considering, it is said, is not one *merely* of *design*, but must rise *from design to a designer*. And here it is that some objections have arisen.

On the one hand it is alleged that the argument is insufficient; and, on the other, that it proves too much, and tends to identify nature with the Deity. But both objections seem to me equally traceable to the same primary confusion of ideas as to the real nature of the inductive inferences, and of the obvious distinction between moral and physical causation. This confusion of ideas pervades the remarks of many otherwise able writers. Thus Coleridge observes,—

"All the so-called demonstrations of a God either prove too little, as that from the order and apparent purpose in nature, or too much, namely, that the world is itself God; or they clandestinely involve the conclusion in the premises . . . as in the postulate of a First Cause."*

Natural theology confessedly "proves too little," because it cannot rise to the metaphysical idea or scriptural representation of God. These stand on quite distinct authority. But "the postulate of a First Cause" is a notion wholly arising from the confusion of ideas just referred to.

Objection to this argument.

The common objection to the argument from design to *a designer*, appears to be of this kind. It is alleged that, to take Paley's well-known instance of the watch, we make our inference directly of a *watchmaker* from obvious comparison with known human works. Even when a person should for the first time witness some work far transcending his own power or knowledge, or anything previously heard of, still he would perceive the analogy with the more ordinary productions of human skill, dif-

* Aids to Reflexion, vol. i. p. 139.

fering only in *degree*, and would infer a contriver and an artist of faculties far higher, but still similar to his own. But the works of nature, it is said, differ from these *in kind;* they are unlike any of our works, and suggest no such analogy of an artificer resembling a human artificer, or differing merely in the extent and degree of his skill.

Apparent want of analogy in the two cases.

In those cases most nearly approaching the nature of human works, such as the varied and endless changes in matter going on in the *laboratory of nature*, the results, even when most analogous to those obtained in human laboratories, yet present no marks of the process or of the means employed, by which to recognise the analogous workman; and in all the grander productions, the incessant evolutions of vegetable and animal life, which no human laboratory can produce,—in the structure of earth and ocean, or the infinite expanse of the heavens and their transcendent mechanism, still further must we be from finding any analogy to the works of man, or by consequence any analogy to a personal individual artificer.

Answered by the real analogy of mind.

But the more just view of the case is that which arises from the consideration that the real evidence is that of *mind* and *intelligence:* for here we have a proper and strict analogy. *Mind* directing the operations of the laboratory or the workshop, is no part of the *visible apparatus,* nor are its operations seen in *themselves*—they are visible only in their *effects;*—and from effects, however dissimilar in magnitude or in kind, yet agreeing in the one grand condition of *order, adjustment,* profound and recondite connexion and dependence, there is the same evidence and outward manifestation of INVISIBLE INTELLIGENCE, as vast and illimitable as the world wherein those manifestations are seen.

Analogy of moral causation and mind.

It is by *analogy* with the exercise of intellect, and the volition, or power of moral causation, of which we are conscious within ourselves, that we speak of the *Supreme Mind* and *Moral Cause* of the universe, of whose operation, order, arrangement, and adaptation, are the external manifestations. *Order* implies what by *analogy* we call *intelligence;* subserviency to an observed end implies intelligence *foreseeing,* which, by analogy, we call *design.**

* On the *analogical* nature of all our modes of speech respecting the Divine Being and attributes, the reader should especially consult the most luminous and philosophical view given by Abp. King in his "Discourse on Predestination," 1709, § 8. Reprinted 1821, with Notes by Archbishop Whately.

Again: nothing but the common confused and mistaken notions as to laws and causes, could give any colour to the assertion that "the argument proves too much," that physical speculations tend to substitute general physical laws in the place of the Deity; and that scientific statements of the conclusions of Natural Theology are nothing but ill-disguised Pantheism.

Invariable laws and causes no argument for Pantheism.

The utter futility of such inferences are at once seen, when the smallest attention is given to the plain distinctions above laid down between "moral" and "physical" causation; and to the proper force of the conclusions from natural science establishing the former by means of the latter.

This distinction obviously points to the *very reverse* of the assertion that physical action is identical with its moral cause; the essential difference and contrast between them is the very point which the whole argument upholds and enforces.

So utterly preposterous, indeed, is the whole idea of Pantheism that the profession of it, if sincere, can but be a mystical fancy, of the most perplexed and unintelligible nature, and involving moral contradictions of the grossest kind; at all

Pantheism absurd.

events, widely alien from any legitimate conclusions of physical science or philosophical speculations based on sound views of causation.

Conclusion. The whole tenor of the preceding argument is directed to show that the inference and assertion of a *Supreme Moral Cause,* distinct from and above nature, results immediately from the recognition of the eternal and universal maintenance of the order of *physical causes,* which are its essential *external manifestations.*

Of the *mode of action or operation* by which the Supreme Moral Cause influences the universal order of physical causes, *we confess our utter ignorance.* But the *evidence* of such operation, where nature exists, can never be lost or interrupted. And in proportion as our more extended researches exhibit these indications more fully and more gloriously displayed, we cannot but believe that our contemplations are more nearly and truly approaching their SOURCE.

ESSAY II.

ON THE

UNITY OR PLURALITY OF WORLDS.

§ I.—THE ARGUMENT CONSIDERED IN A PHYSICAL AND PHILOSOPHICAL POINT OF VIEW.

LORD ROSSE'S TELESCOPE.

AMONG the endless topics of human inquiry and controversy, we cannot but observe that those often become the most popular which might have been antecedently pronounced the most unlikely to attract notice: and subjects the most remote from those which ordinarily engross the attention, involve the interests or excite the passions of mankind, and even those of the most imaginary and conjectural character, have often called forth the most earnest dispute.

Introductory remarks. Propensity to discussion of remote and imaginary topics.

Often, too, it happens that a particular opinion has for a long time been conventionally adopted, or

Love of novelty.

acquiesced in—perhaps with little inquiry or thought among the many—until on a sudden some writer, more bold or more ambitious than his predecessors, discovers a dubious point on which at least a plausible argument may be raised, calling in question the received belief in which the public mind has hitherto reposed. And then, however abstract, or however trivial the subject may be, prepossessions are aroused, and the question is immediately raised into serious importance, and a controversy stirred up whose vehemence is often just in the inverse ratio of the real value or clear evidence of the point at issue.

Interest felt in conjectural speculation.

Men who take comparatively little interest in the tangible details of real attainable science, feel stimulated by the desire to penetrate those recesses where all is obscure, and *certain* knowledge unattainable. There is indeed a line of demarcation, nowhere more clearly drawn than in the positive mathematico-physical sciences, between the known and unknown; yet on the frontier there lies a region on which sufficient light is shed to permit our curiosity safely to indulge in short incursions, while there is darkness beyond which we may people with unsubstantial forms and shadows to the full satisfaction of our wildest imaginations. And this is pre-eminently the case in the vast expanse opened to our view by astronomy; and when readers and inquirers are tired with the dry discussions of periods and distances, and with calculations of masses and eccentricities, they naturally fly for relief to the more grateful occupation of guessing at the probable near aspect of the surfaces of the heavenly bodies —imagining them worlds like our own, and fancying the possible nature of the creatures who may be their inhabitants. And as such subjects are of course open to unlimited conjecture of every kind, so are they liable to become the battle-field of interminable dispute: interminable, perhaps, some may say, in proportion as the hosts of objections raised on one side, and of replies on the other, are all equally unsubstantial; and like the aërial combatants of the poet of impassive and indestructible nature, reuniting as soon as cloven asunder, so that, after all is over, either side may with equal confidence claim the victory, and be ready to begin the fight anew.

Popular belief in a plurality of

Yet, to the question of the existence of inhabitants in other worlds (so impossible really to inhabited worlds.

answer decisively) there seems to have prevailed a general disposition to give in some sort an affirmative reply; not, perhaps, founded always on very scientific grounds; but undoubtedly the common expression of popular books has been favourable to the idea, and perhaps most educated persons would hardly have hesitated to admit its general probability.

Views of the subject in former times.

In fact, no sooner was the true planetary system generally received, than it became an obvious topic with writers who were engaged in recommending it to public notice, to invest the dry details of planetary astronomy with the interest excited by speculation on the possibility of intelligent beings inhabiting those distant worlds. And when men had become reconciled to the paradox of *antipodes*, it was not a much greater difficulty to concede *Lunarians, Jovians, Saturnians,* and the rest. And thus the general idea of inhabited worlds, under various aspects and with various objects, has been alike upheld by philosophers and poets—by divines and popular essayists. Hinted at even by Newton and Huyghens—recommended to popular acceptance by the elegant discussions of Fontenelle—reasoned upon theologically by Derham and Bentley, it passed into a matter of popular credence; and was often appealed to even in later times, both as a fair philosophic hypothesis and as a worthy religious contemplation by Herschel and Paley— by Lardner and Chalmers, and by Dr. Whewell in his Bridgewater Treatise.

Discussion at the present day.

At the present day, it hardly needs to be remarked, on this subject a contest has been keenly carried on between two highly talented disputants, whose publications* have called forth an unusual amount of popular attention; and the question of the *plurality of inhabited worlds,* which had altogether slumbered in public interest since the days of Fontenelle, has once more started into life, and occupied the public mind; and after the public scepticism had been at length satisfied as to the *actual rotation* of the earth by the pendulum experiment of M. Foucault, this discussion took its place as a fashionable topic, and commended to the public favour the Copernican hypothesis; on the one side, as perfectly

* 1. "The Plurality of Words," an Essay: 1853. 2. "A Dialogue on the Plurality of Worlds," being a Supplement to the Essay, &c. 3. "More Worlds than One; the Creed of the Philosopher, and the Hope of the Christian," by Sir David Brewster, K.H., D.C.L., F.R.S., V.P.R.S. Edin., and Associate of the Institute of France, 1854.

safe, when carefully divested of the dangerous adjunct of imagining inhabitants in the planets rivalling man in dignity and spiritual privileges; on the other, as eminently orthodox, if the planets be believed to be tenanted by such inhabitants, whose existence is even demanded by religious considerations, and made at once a leading point in "the creed of the philosopher and the hope of the Christian."

In the instance of the present controversy, whatever opinion may be formed as to the issue or the merits of the question, there can be but one as to the ability with which each disputant has conducted his argument; and especially as to the ingenuity which the anonymous writer has evinced in maintaining what must seem in many respects somewhat paradoxical theories, the more freely thrown out under his assumed disguise.

Various ideas broached of the inhabitants of planets.

The literature of such a question is not without its curiosities; and both the authors referred to have contributed some account of the varied opinions which have been broached on the subject. The author of "The Plurality" displays his stores of erudition on this head in furnishing several singular exemplifications.

Some of the ancient Stoics, as we learn from the ridicule cast on them by Lactantius, ascribed inhabitants to the moon. Plutarch, in his curious dialogue "on the face which appears in the moon's orb,"* gives arguments for and against the moon being inhabited. Lucian indulges in the same fancies; but probably neither with any settled conviction of their truth. Nicolas of Cus, who asserted the Heliocentric system before Copernicus, discusses in some measure the nature of the solar and lunar beings, and makes the former more intellectual, more clear and illuminated, than those of the moon, who are "magis lunatici," as in the earth they are "magis materiales et grossi."

These were followed by the strange assertions of Giordano Bruno, who declares not only for a plurality of worlds, but that the earth is inhabited in its interior as well as its exterior; and Wilkins, who laboured to prove not only that the moon is inhabited, but that we need not despair of being able to visit the inhabitants. Kepler expressly argued in favour of Lunarians: "Consentaneum est esse in lunâ viventes creaturas."

* Supplement, p. 36.

Even down to later times, imagination has revelled in devising the kind of existences which may inhabit the planets. Fontenelle assigned corporeal beings very like ourselves at least to the nearer planets; creatures of extreme vivacity to Mercury; of voluptuous and ardent natures to Venus; of more robust and manlike character to Mars; while to Jupiter and Saturn beings of dull and torpid constitution were given. Sir H. Davy, in his vision, saw creatures of the most marvellous structure, with membranous bodies and strangely convoluted elephantine probosces as organs of sense and intelligence, gifted with far higher intellectual capacities than the men of this earth, inhabiting Saturn; to which we may add, that a very ingenious and scientific poet has recently depicted beings in the moon with an *internal* body and an *external* soul.*

It is not, however, to any such familiar acquaintance with other worlds that our immediate inquiry is to be understood to aspire. The actual point in question is not the most plausible image we can

* "Love in the Moon," a poem, with Remarks on that Luminary, by Patrick Scott. London, 1853. See p. 24.

conjure up of the nature and appearance of lunarians or planetarians, but what is the most proper and philosophical view we can take of the general question, *Are other worlds besides our own* PROBABLY *the seats of intellectual, moral, and spiritual life?* Is it probable from concurrent circumstances that our globe is so far a peculiarly conditioned portion of the whole creation as to be the only one privileged in this respect? or are not others, or all others perhaps, equally, or even more, elevated in their destination as seats of life?

Statement of the question.

The question is one which most persons at all given to contemplate the phenomena of the heavens have been always prone to ask, though perhaps with little serious conviction of the possibility of answering it with any reasonable approach to certainty. But it takes a more precise form, and the real nature of the inquiry is more distinctly indicated the more we consider the actual conditions of the sidereal world. The question, probably, first arises with respect to those bodies nearest to us, and which most resemble our earth. From the obvious general resemblances between our globe and those others of the solar system which might be regarded

General analogies.

as members of the same family in respect to form, motion, subjection to the laws of gravitation, diffusion of light and heat, in some instances the presence of moons or rings affording auxiliary illumination, in others of atmospheres, clouds, and therefore water, of mountains and valleys, or even of supposed continents and oceans, there might seem to be an easy transition to the belief in intelligent inhabitants, bearing more or less resemblance to ourselves.

The inquiry perhaps is first made,—Is our companion the moon inhabited? It then extends to the other planets, and as it extends, it may seem surrounded with more difficulty; are we to include the sun? or is not his nature so different as to render such an inquiry unreasonable? Are comets likely to be inhabited? still more the fixed stars? but they probably are suns; they may be the centres of planetary systems — of worlds; why not of animal life, or even of intellectual and moral life?

But there are yet more distant bodies, the systems of nebulæ and clusters, to which the same questions may apply; and thus the overwhelming magnitude of the inquiry makes us the more alive to its difficulties, and we feel ourselves lost in the vastness of the conceptions it inspires.

Influence of modern discoveries on the question.

At the present day, from more accurate investigations, aided by the recent improvements in telescopes, the actual structure and physical conditions of the planetary bodies of our world, as well as those belonging to the more immensely distant and vast sidereal systems, have been better known to astronomers, and correct ideas respecting them more familiarly diffused in popular information. Hence it becomes a point of inquiry whether these accessions to our knowledge have been such as in any way to affect the previously received notions as to the existence of organised life in the heavenly bodies.

To follow up this inquiry is the professed object in a large portion of the discussion now brought before the public; and it is apparent, that while the modern discoveries generally have confirmed and extended the analogies of planetary and stellar systems, they have also disclosed many particulars which may require us to modify our notions in detail as to the conditions of their existence; while geological research has not been without its bearing on the question of their structure, nor the various cosmical and cosmogonical theories altogether uninfluenced by the latest discoveries of nebular astronomy.

It was thus but fair and reasonable that the question should at the present day undergo a renewed discussion; and whatever opinion may be formed as to the precise result to which the present controversy may tend, it will, probably, on all hands be allowed that it has not been unproductive in bringing more prominently forward many of the most interesting facts and conclusions respecting the structure and conditions of the heavenly bodies, and at the very least putting the public mind more fully in possession of those data which are necessary for carrying out any more imaginative speculations on reasonable grounds.

The question a point for philosophical conjecture.

In the former Essay* reference has been made to the nature and grounds of *Philosophical Conjecture*, and the place which it may legitimately and usefully hold among the speculations of science. Those

* See Essay, I. § iii. p. 89.

remarks may perhaps find an application in reference to such questions as that now before us.

Viewed simply as a question of philosophical conjecture, or rational probability, without reference to any ulterior consideration, the argument must be based on an extension of *inductive analogies*, a generalisation (so far as we can legitimately pursue it) upon the acknowledged relations of animated existence with physical conditions and cosmical arrangements adapted to it. And it is in this point of view that we must, in the first instance, proceed to consider it. At the same time, so numerous are the points of relation between the simple question of probability as to the fact of inhabitants in the heavenly bodies, and various collateral topics of higher interest, that the larger portion of the discussion, as taken up by the disputants already referred to, is in fact chiefly occupied by these collateral topics, to which the more simple question is manifestly regarded as subordinate: and it is probable that the public has been induced to feel an interest in the subject more from a reference to such ulterior considerations than from the intrinsic attractions of the primary question itself.

Review of the argument necessary.

But notwithstanding the ample discussion which this subject has received from two such eminent disputants, it is still, in my opinion, left by them in an unsatisfactory state. Not so much in regard to the mere question itself, which ever must remain in uncertainty, as with respect to a just appreciation of the true grounds on which the discussion of it should be taken up, as well as of the bearing and influence which it may have upon those higher contemplations with which both these writers (though in opposite ways) have combined it. It is, then, in this general point of view, and in its connexion with other topics of philosophical inquiry, rather than as to the mere details of astronomy, that it is here proposed to treat the subject. Yet to a few of those details some attention must be paid in the first instance.

Connexion with past history of the world.

The question as to the probable *present* habitable condition of the planets or the existence of intelligent beings upon them, is closely connected with the *past* history of the system. And the discussion of inhabited worlds has been much mixed up with that of the process of transitions through which they may have passed from an original nebulous or vaporous state to their existing condition.

The nebular theory.

The "nebular theory," as it is termed, of the origin of *our planetary system* is totally distinct from the phenomena of the *sidereal nebulæ* with which, nevertheless, it is often confounded. When the former theory was broached by Laplace, he observed, indeed, that it received some confirmation by analogy from the then recent discoveries of the sidereal nebulæ by the elder Herschel.

Sidereal nebulæ.

That highly distinguished astronomer had observed with his powerful reflectors and minutely described the forms and characters of a great number of those nebulæ, and thus there were supposed to be furnished so many actually existing instances of what had as yet been only a hypothetical speculation. Nebulous matter had been assumed to have existed in the solar system; here were cases of such matter really existing in the sidereal heavens. Moreover, there were great diversities of form and species of such matter. Some were mere faint patches of dull light: others exhibited something like a nucleus, or brighter centre: others a distinct star surrounded with nebulous matter: others various combinations of stellar and nebulous appearances, often presenting irregular portions, and many of variously formed globular or elliptical structure. Hence he was naturally led to the conjecture that these might be only gradual and *progressive stages of formation* from mere diffuse nebulous matter up to condensed stars, or solid masses.

Sidereal clusters.

And the analogy was further carried out when, besides the nebulæ properly so called, there were found a number of other bodies which to lower powers appeared like nebulæ, yet with telescopes of higher capacity were *resolved into clusters of separate stars.* This, then, seemed to be the ultimate stage of their progressive evolution, originating out of mere diffuse cosmical nebulosity, by degrees condensed towards various centres, and at length absorbing all the matter into distinct bright stellar bodies clustered together—that is, clustered *to our eyes,* but really at amazing distances from each other—forming vast but separate groups in the heavens: of which groups the whole of the fixed stars (ordinarily so termed as distinct from the clusters and nebulæ), extending laterally, and seen in thick perspective to form the Milky Way, are believed to constitute only one, perhaps very subordinate, component part or cluster; all the members of which are self-luminous, or suns;—one among which is our own sun, with his little attendant planetary system, as invisible to them as any systems they may possibly possess must be to us.

Relation of nebulæ to the solar system.

We might well stop to expatiate on the magnificent scene thus presented to our contemplation, as both of the writers referred to have done with so much eloquent effect. But to return to the immediate argument; it becomes important (with reference to a common confusion of ideas) to dwell upon the consideration that the analogy of the sun and his system is *not* with *clusters* or nebulæ, but with *their integrant stars.* And those who expect either to confirm or to refute any supposed relations or conditions of our planetary system with what may be observed or imagined in the sidereal nebulæ are altogether on a wrong course.

Resolution of nebulæ by powerful telescopes.

Now the nebulæ, as originally observed in the northern hemisphere by the elder, and in the southern by the younger Herschel, have been since, to a considerable extent, re-examined in the northern with the vastly more powerful telescope of Lord Rosse; and, as the former observations had greatly altered our conceptions of such nebulæ as had been previously examined by Messier and others with very inferior powers, and had resolved into separate stars some which had previously been regarded as purely nebulous, so the more gigantic powers now applied not only presented many of the familiar nebulæ under aspects in which their old features could hardly be recognised, but succeeded in *resolving* into *clusters* of *stars* many objects before deemed incapable of such resolution. And it may be well here to observe, that when the author of "The Plurality" describes these resolutions as into *patches* and *dots* of light, seeming (if I understand him rightly) rather to discountenance the notion of their strictly *stellar* nature, in point of fact, I am able to state, on the authority of those who have actually seen them in Lord Rosse's instrument, that the appearance is perfectly and brilliantly that of *stars;* distinct effulgent points of no sensible magnitude, and of whose stellar nature no doubt could remain on the mind of the observer.

Here, then, an argument has been raised:—As successive portions of nebulæ become resolvable when we apply successively higher powers, so it is inferred we may reasonably extend this argument, and fairly expect that this would always continue, as still higher powers might come to be employed; and thus there would be no limit, and, in fact, all sidereal nebulous appearances would thus be shown to be most probably nothing but starry clusters, appearing nebulous simply in consequence of their enormous distance. Hence, it is said all theories founded on such conceptions of nebulous matter must be given up; and hence the opponents of such theories have enjoyed a triumph in which they have been often prone to indulge feelings not apparently much in character with so purely abstract and hypothetical a speculation, on which, nevertheless, opinions on either side have been maintained with a degree of vehemence little to have been expected from the nature of the subject. To this, however, there is opposed one remarkable fact — viz., that in all the instances examined by these extremely high powers, wherever parts of nebulæ not before resolved have been so resolved, there have been also brought to light numberless new portions of the same mass of a character apparently quite nebulous, of extreme tenuity, bending about in delicate films in the most capricious shapes. It has hence been an inference

Whether any limit to such resolution?

on the other side, that with widely ramified stellar clusters there are always associated large masses of yet unformed cosmical matter.

Some nebulæ real: not merely optical.

This perpetual disclosure of new unformed nebulous filaments and appendages to central stellar clusters has been associated by the author of "The Plurality" with some other considerations arising out of the phenomena of certain other nebulous appearances in various parts of the heavens, especially those singular and well-known bodies the "Magellanic Clouds," so minutely and graphically described by Sir J. Herschel in his "Observations in the Southern Hemisphere." These present, in remarkable juxtaposition, examples of almost every form of stellar and nebular phenomena. All which the author considers as supporting the conclusion, that these nebulous appearances are actually and properly such, and differ from clusters of stars, not merely in semblance from the optical effect of distance, but in their own nature as real aggregations of diffuse cosmical matter of some kind.*

As to the probable distances of this class of nebulæ,

* Essay on the Plurality, &c., p. 118.

Probable distances of some nebulæ.

it has been observed, that the truly nebulous portions are evidently *physically* connected with the bright isolated stars scattered among them; and these we have no reason to suppose more distant, on the average, than other stars. Lord Rosse, in his address to the Royal Society, 1853, observes, "In certain nebulæ, stars are so peculiarly situated that we can scarcely doubt their connexion with the nebular system in which we see them, and some of these stars are as bright as some of the stars known to be physically double; as bright even as some of the stars which the latest Pulkowa observations have shown to have sensible parallax, and whose distance is therefore approximately known." This agrees with the argument of Sir W. Herschel*, on the principle of the visibility of stars in more powerful telescopes, being a measure of their distance, he calculates, that, as stars are visible to the naked eye up to a certain order of magnitudes, that is, distances, so telescopes bringing into view more orders, have their space-penetrating powers numerically assigned on that scale. He ob-

* Phil. Trans. 1818.

served, with the highest powers, a star which was itself of the twelfth magnitude, still surrounded by a nebulous haze. If the nebular part really consists of stars, only not separable by the telescope from the effect of distance, then the central star must be of such enormous magnitude, in comparison, as would be at variance with all analogy; and hence he infers that the appendage is truly nebulous.

The great nebula in Orion remained unresolved by all telescopes, even Lord Rosse's smaller one; and when, in 1846, the large reflector was applied, though much of it was resolved, or gave appearance of "resolvability," yet it would seem that some portions still retained a nebulous appearance.*

Such considerations seem to indicate a probability that some of these really nebulous masses may be not more remote than some of the single stars which compose our own cluster.

Question of proper motion.

Some real test would probably be supplied, if any of these nebular bodies should be found to possess *proper motion,* such as so many of the fixed stars have been ascertained to possess, and which

* See a Short Letter addressed by Lord Rosse to Prof. Nichol, published in his "System of the World," p. 55.

stand out as residual phenomena, after all deduction for the ordinary corrections, and have been shown principally to arise from the real motion of the solar system in space. Hence the existence or amount of *proper motion* is, generally speaking, something like a measure of *distance.* If then, other clusters and nebulæ, not connected with single stars, should show considerable proper motion of such a kind as accords with the motion of the solar system, it would be a proof of their proximity. But until such proof has been given, it is obviously a premature generalisation to assert of all the nebulæ and clusters *universally*, that they are not more remote than the stars, because *some* of them may be so. If the theory, that *all* nebulæ are only remote clusters is to be abandoned, it can only legitimately be exchanged for the assertion, that we must distinguish nebulæ into two classes, those which are *near*, and are *truly* nebulous, and those which are *remote* and only *apparently* so.

Question as to revolutions of nebulæ about their centres.

Whether the parts of clusters and nebulæ may be in motion round any point, is a question which has not hitherto been much inquired into. Some theoretical remarks have been thrown out by Sir J.

Herschel *, and some doubts expressed by others as to the possibility of their preserving equlibrium without a rotatory motion, or as to the possibility of the law of the inverse square of the distance being sufficient to account for their phenomena. It is, however, obvious, that if there were the most rapid revolution in any portions of these systems, they may be at too immense a distance to enable us to detect it till after centuries of observation. The sail of a windmill or a railway carriage sweeps past the spectator close to it with lightning speed; seen at the distance of some miles, it seems to revolve with extreme slowness, or to creep along, in proportion to the diminished *angular* visual space passed through in the same time. A distant planet moving with the velocity of several thousand miles in a minute, presents no sensible motion even to the astronomer, with the nicest instruments, except by comparison of observations after considerable intervals. Double stars describe enormous orbits with velocities of proportionate amount, whose motions are not to be detected but

* See Outlines, p. 866.

by the most delicate measures at distant epochs. How much more, then, may the infinitely more remote components of a starry cluster seem to be at rest even though they may really be whirled in a second through inconceivably vast regions of space about their centre of gravity?

Spiral forms of nebulæ.

But there is a still more striking point which has been duly commented upon by both the writers before us — that singular feature pervading so many of Lord Rosse's nebulæ, that they appear in the form *of spiral convolutions of filmy nebulous matter, tending to a central nucleus or star.* And an analogy is immediately suggested with a revolving *system,* — with bodies urged towards a centre in contracting spirals, because the orbits they would have described round that centre are continually compressed by the action of a *dense retarding medium* through which they move.

Probable revolutions in spirals.

Lord Rosse* distinctly states his belief in the probability of *motion* in the spiral nebulæ: — "If we see a system with a distinct spiral arrangement, all analogy leads us to conclude that there has been motion; and that, if there has been motion, it

* Address to the Royal Society, 1853, p. 7.

still continues. The apparent motion is probably very slow, owing to the immense distance of the nebulæ."

The filmy spirals of the nebulæ *may* doubtless be long strings and series of separate stellar bodies, all performing revolutions, whether in a retarding medium, or by virtue of the action of central forces acting by other laws than the inverse square; such, for instance, as the inverse *cube*, which, as is well known, Newton proved would cause a revolution in spirals. Our sun, with his attendant system, has been shown to be in motion in space; and probably revolving round a point assignable among the stars, and which some have fixed in or near the Pleiades. But if in *our system* there be supposed any analogy with the spiral nebulæ, it should be remembered it is *not* with the motions of our planets or of Encke's comet and the retarding medium which it discloses, but with the cosmical revolution of our *whole system*, as part of *a cluster:* and to make out the analogy, it would be necessary to inquire whether it may possibly be the case, that other members of the solar cluster besides our planetary system, partake in a *sidereal* motion, and whether our milky way and little stellar group may be seen by the astronomers in *other clusters*, assuming the form of spiral wreaths wheeling round the Pleiades, and destined ultimately there to condense, and console the survivors for their lost sister.

Too limited views of the extent of the sidereal heavens.

The language employed in some parts of the speculations now adverted to, seems rather of a nature calculated to impose limits on our ideas of the immensity of creation, and thus to check one of the most sublime contemplations in which science has hitherto given us the privilege of indulging. The nebulæ can no more properly be called the "outskirts of creation," than the nearest planets. "What is there on the other side of the stars?" is a question which a child may ask, but which the philosopher can only answer by alleging the fact of increasing numbers of sidereal masses becoming successively visible as our telescopes penetrate further into space. To take up a theory which tempts us to put limits on the extent to which new worlds may become visible to us if our telescopic powers were conformably increased, would be in direct opposition to those suggestions to which all analogy points; to limit the immensity of creation, and to attempt to model the plan of the universe to our own narrow dogmas. A microscopic animalcule on the sea shore might as well conclude that the whole universe consisted of nothing but those enormous masses, the grains of sand, which are all that he could see around him, and as well deny the existence of the land and the ocean, with their varied inhabitants, because beyond his limited vision.

The speculations in which the true philosopher may indulge on such a subject, guarded as they ought to be by a becoming caution, will yet tend continually to the increasing conviction at once of the boundless extent of the universe, and of the order and harmony which he is assured must pervade every part of it.

Nebular question as bearing on that of "inhabitants."

But in the present instance the question of nebulæ and the nebular hypothesis has been taken up not simply on its own merits, but as bearing, or supposed to bear, on the question of the *habitable condition* of these cosmical bodies or regions.

The resolvability of all nebulæ into separate stars and worlds is upheld as favourable to the notion of abodes for rational beings: and the existence of nebulous matter is discountenanced by those who are anxious to people every point of the resolved masses with animated existence. While, on the other hand, those who would pronounce the whole material world to be a vast and eternal "waste," excepting one small and favoured speck, find the hypothesis of nebular expansion through many enormous tracts of space more congenial to their ideas.

But it is difficult to perceive any very close or necessary connexion between the two questions. If, indeed, true nebulous matter be shown to exist in some instances, the real "star dust," or "world mist" of Humboldt, it certainly would not seem a very suitable dwelling-place for any kind of organised inhabitants: yet this might be merely because such portions of the system would not be as yet in a stage of development in which we could expect or imagine life to have begun. But if this were true in *some* parts of the universe, surely it would not impugn the existence of inhabitants in *other* portions of the system.

In more advanced stages of consolidation, the progress of life or the preparation for it *might* be going on; and in the regularly formed worlds which may have been long ago completed, it *might* exist in full development notwithstanding that in other parts of the system as yet in a state of nebulosity, we might doubt or deny its existence. If we chose to argue the question whether a consolidated globe be a necessary condition for the origination of life in any form, but more especially of its higher forms, and whether intellectual and moral faculties are to be restricted to higher forms of organisation, or to any, these are distinct inquiries.

The fixed stars.

But the fixed stars propely so called—the members of our own cluster—some of which are revolving systems, binary, ternary or even multiple;—the distances of some of which are actually within the reach of our measures, and in comparison with whose distances the diameter of the earth's orbit is not an *absolutely* insensible point—are they inhabited? They consist of matter kindred with that of our world; they are subject to the marvellous affinity of the force of gravitation; they emit light, that same agent which affects our eyes, and which terrestrial bodies emit. Whether we admit the inconceivable idea of a universal ether in which these remote bodies excite vibrations capable of ultimately reaching us, or the equally or more inconceivable idea of material molecules darted forth from these same bodies and actually projected into our eyes—and *one* of the two we *must* admit—these and many more considerations show their kindred nature. But being self-luminous, are they not suns? Must they not be, like our sun, the sources of heat also, and therefore probably intensely hot? If so, can they be inhabited?

Self-luminous, or suns.

That the stars are self-luminous, and therefore (it might be argued) probably consist of matter in a state of incandescence or combustion of some kind, is the strongest fact against their being inhabited. This however is no more than the same argument relative to the sun, and, like it, susceptible of the same answer from conjecture as to the existence of an exterior photosphere. But this will be considered in the sequel.

It must also be recollected that the revolutions of binary or multiple stars are not analogous to those of our planetary system, but are the revolutions of two or more suns with their attendant systems (if any) about their centre of gravity.

The orbits of some of the double stars are indeed

Revolving systems of double stars.

not greater than those of the exterior planets of our system, though their periods are rather greater (evincing consequently a less attraction between the two stars, or what is the same thing a joint mass less than that of the sun); yet small as these orbits are compared with that which our sun is describing round the Pleiades (or whatever the central point may be), the analogy would seem rather to be with this last orbit, than with our planetary orbits. Whether in these instances the two mutually revolving suns may be attended each with his system of planets is a point which, doubtless, no existing observations tend to clear up. But it is one which theory pronounces unlikely, on the ground of the too great proximity of the rival attractions of the two suns, which would make the maintenance of such planetary systems impossible, unless we suppose all the members of each system to lie extremely close to its sun.* There is, however, no reason whatever why they should not be thus constituted; nor why non-luminous planets thus arranged should be without inhabitants.

But even were the *double* stars suns, destitute of

* See Herschel's Outlines of Astronomy, p. 847.

planetary systems, there is no assignable reason why the *single* stars (constituting so infinitely greater a proportion) should not be thus attended: it being obvious that no conceivable telescopic power could ever discover such worlds to us. While some of the *nebulous stars* might clearly present the rudiments of such systems yet to be formed.

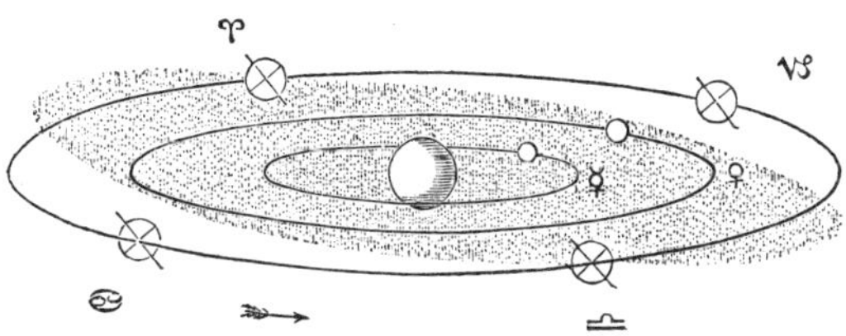

Zodiacal Light-mass.

When from these remote systems we turn to that with which we are more intimately connected, we certainly recognise the existence of portions of *true nebulous matter* in several parts of our planetary world. We find it in *comets*—in the *zodiacal luminous mass,* occupying nearly all the space within the earth's orbit—in the *haze* surrounding more than one of the small planets, and most probably in those as yet little known forms of cosmical matter, revolving under the influence of gravitation, which

Nebulous portions of solar system.

give rise to meteors, and especially the periodical star-showers.

To give some idea of the magnitude of such masses, it may suffice to mention Encke's comet, which at a distance from the sun equal to that of the earth, occupied with its nebulosity a sphere greater than that of the *whole* moon's orbit; and the comet of

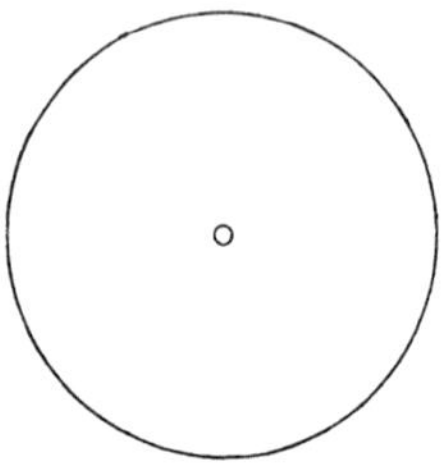

EARTH, AND MOON'S ORBIT.

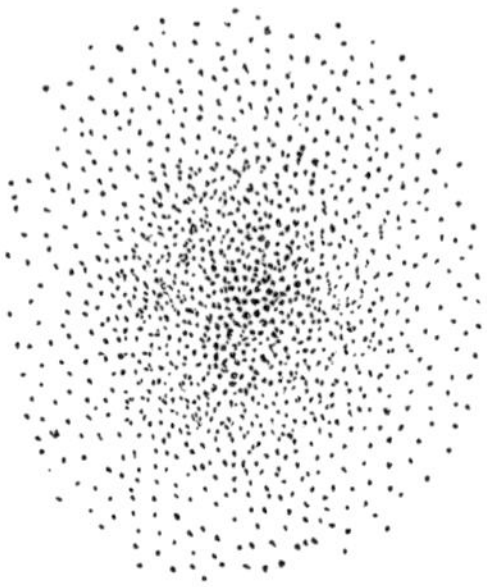

ENCKE'S COMET, (Relative Size.)

1843, whose tail according to Boguslawski equalled in length the *entire distance from the sun to Jupiter*, or 494,000,000 miles, with an average breadth of perhaps 500,000 miles. And when we further take into account the multitudes of comets, so great as to justify the assertion of Kepler, that the universe is full of them, we shall have little remaining doubt as to the plentiful existence of true nebulous matter.

Indeed, the late Mr. Baily, than whom no man was less likely to be led astray by fanciful hypotheses, did not hesitate to speak of comets, as being detached fragments and remains of the original nebulosity; which "want of mass has saved from the extreme condensation which the planets have experienced;" as having been originally projected in parabolas, and then, perhaps, from the resistance of the uncondensed remaining nebulous matter with which they were surrounded, compelled to revolve in limited and re-entering orbits,—and as forming "a link between the present and past states of our system, showing in their obedience to the law of gravitation a presumption that that law has been unaltered since the system was nebulous, and now holds good in the nebulæ which remain unchanged."*

Nebular origin of our system.

The existence of such unformed matter adhering to our system has been appealed to as at least affording presumptive evidence bearing on the probable origin of that system; and while one party has been led to support the nebular theory of Laplace, perhaps to an undue extent, and to find in

* Address to the Royal Astron. Society, 1837. Notices, vol. v. p. 51.

it a solution of the entire creation of our system, in some instances carried out by attempts at even *numerical* verification, which have been shown to be erroneous,—another party has with equally unphilosophical vehemence and warmth denounced that theory as replete with every form of mischievous error, and has indulged in misplaced feelings of triumph when they fancied they saw its downfall in the detected errors of M. Comte's computations, or the resolution of the sidereal nebulæ in Lord Rosse's telescope.

It is no part of the present object to assert or to defend the nebular theory, except on the general ground that it is a perfectly *legitimate* kind of *conjecture.* One objection, however, may be noticed in passing, because it has been sometimes referred to as destructive of that theory. It arises, however, from a mere oversight. A body *rotating on its axis* in a dense medium will have its motion simply *retarded;* a body *revolving* freely in an *orbit* in the same medium will have its *orbit contracted* and its *motion* consequently *accelerated.* This distinction has apparently been lost sight of by some who raise objections to the nebular theory, on the ground that

An objection answered.

the times of the rotation of the sun and of the revolutions of the planets are so different. They must necessarily become so under the conditions supposed, though originally one rotating mass. As soon as a planet was separated, it was transferred from the dominion of the one law to that of the other.

Rational claims of the nebular theory.

The nebular theory of the solar system, soberly understood, is a philosophical conception worthy of the subject which it illustrates: starting from the fact of central heat in the earth, and the indications of it in the spherical forms of the other planets, we are unavoidably carried back to a period when all was in fusion; and thence to a period when all was vapour or nebulosity, out of which by successive cooling and condensation, and the rotatory motions ensuing upon the rush of matter towards the centre, the existing system *may* have been developed according to regular and uniform laws; and which is so far a rational and consistent *conjecture* (for it can be no more), eminently conformable to the grand principle of cosmical unity and order. The truly philosophical advocate of such a theory, following the track of inductive analogy, might not be disposed to assign organised inhabitants to any of the bodies so formed, till after immense periods of cooling and consolidation. But he would be led into no dogmatising on the subject, and would simply call on us to be guided by the analogies suggested by what we know, subject to the condition that in the infinity of what we do not know, equally grand principles of order and unity must prevail: principles and laws not necessarily the same as those with which we are acquainted, yet equally invariable under the conditions to which they are adjusted.

Common origin of all planetary matter.

But if the common origin of the planets and the sun, from one primary nebulous mass be admitted, this further consideration is forced upon us; viz., that as they were thus all parts of the same material mass, that mass must have contained, mixed up in it, all the elements of every possible product of nature, organic or inorganic, and the germs of all vitality, even to its *highest* forms, *in so far* as they partake of an animal nature; and we may therefore suppose in all the planets the same inherent capacity for having life evolved in them from its lowest up to its highest forms.

Present condition of the planetary system.

If from the conjectural past history of our system we proceed to consider its *actual* condition in reference to the question of inhabitants in the various bodies composing it, we must look carefully to facts, and by the writers who have treated the subject on either side, we are sent back to our works on astronomy, to the ascertained data of observation, and the inferences from them, to consider the magnitudes and distances of the planets, the proportions of light and heat they receive, the variations of their seasons and lengths of their years; the satellites furnished to so many of them, and the rings to one; we are shown well-known calculations of the force of gravity at their surfaces, the weights with which beings placed there would be pressed, the known density of the materials of which they are composed, the presence or absence of atmospheres; the former being admitted, on all hands, in every planet except our moon, and even there still a question with some. From all these and the like data, fair conclusions may be drawn as to their *capability* of sustaining organised inhabitants, and thus we may be legitimately led to *conjectures* as to the probability of rational or moral beings tenanting their surfaces.

Diversity in inhabitants, if any.

Whatever may be the opinion entertained on these points, it will be on all hands admitted that we must suppose, at the least, great diversity in the nature of the possible inhabitants of the different planets, corresponding to the known diversities of conditions subsisting in them. But the inquiry mainly refers to the question, whether these differences must be so great as to preclude all idea of a nature *analogous* to our own, in a physical point of view; it would be of course easy to grant metaphysical entities or spiritual existences of an unknown kind, wholly different; but this is not the question before us.

We are acquainted with moral and intellectual life in finite beings only as connected with a material organisation; we see our own world suited to be the dwelling-place of such organisation; we see other worlds around us presenting many external points of analogy—are they suited to be the dwelling-places of beings in any way *analogous?* The question rather is, whether there are any positive grounds for supposing the diversities so great as to destroy all idea of any common or kindred nature in such inhabitants, and thus practically to put them so entirely out of the category of beings within any range of sympathy or connexion with ourselves, as would be virtually to deny their existence.

Difficulties to be fairly considered.

The object of this essay is not controversial; it is not, therefore, intended to go into the questions of detail raised by either of the disputants on the subject of the habitable nature of the planets. It is my object to look at this question rather in its more general aspect, than as referring to particular planets, and in relation to the broader argument applying to all. There are, however (as we have already noticed), a variety of considerations to be taken into account; with all analogy in favour of the *possibility* of inhabitants in the planets, it is still little more than a possibility. If there were no arguments of another kind to oppose, it might amount to *probability*. But there are some such opposing arguments, which must be carefully noticed, such as those derived from the state of our moon, the geological history of our earth, and the condition of the sun, the asteroids and comets, to all which we shall refer in due order.

The sun.

As to the sun himself: the ingenious speculations of the elder Herschel and Arago have assigned him a race of inhabitants very like ourselves, living upon the solid globe or nucleus, over which is spread the resplendent atmosphere or photosphere, the source of those rays which convey both light and heat, and which, it is conceived, may emanate from the envelope without affecting the central body.

Nature of the solar heat.

Indeed, on this point there is one consideration often not sufficiently attended to. The solar heat is entirely of a peculiar nature, unlike that which emanates from a terrestrial hot body simply cooling or radiating its heat. The solar heat is not derived from the mere cooling of the sun, but is conveyed, as it were, *in* the rays of light, as a *vehicle*, and *never* becomes *sensible as heat* till the *light* is absorbed. It is, therefore, probable that these rays may owe their extrication from the sun to *some other cause than elevation of temperature*. It is an effect elicited or produced by the action of certain rays, which are no more properly rays of heat than a galvanic current can be called a current of heat, because, when stopped, it excites heat. The solar rays pass freely not only through empty space, but even through air and all *transparent* media, *without heating* them; they never excite heat till they are impeded by a solid, or at least an opaque body.

"The temperature of space" is a term which some philosophers use, probably meaning the proper temperature of *some medium* diffused through the celestial spaces, and which is independent of the sun's radiation. This proper temperature is supposed by some to be extremely low; at any rate, it manifestly depends on the degree of this temperature to what extent any planet shall *retain* the heat imparted to it by the sun, and the loss of heat will be greatly increased in proportion as their surfaces possess high radiating powers.

Temperature of the planets.

The time of rotation of the planets, again, is a material element in modifying the degree of heat they receive from the sun, from the comparative rapidity with which points on their surfaces pass from under the heating rays of the sun. But this, again, must be greatly modified by the individual temperatures of the planets arising from internal heat, the remains of their primeval high temperature, and dependent on their rate of cooling, which may be different in the different planets, from a state of primeval fusion. Some rough calculation, perhaps, might be instituted from their known densities and magnitudes which might give an idea as to the possible relative temperatures, or states of consolidation, which they may at the present time have simultaneously reached, supposing them to have commenced from a common period of fusion.

On such points, indeed, we have few data; but they are clearly essential to be taken into account before we can pronounce on the actual temperature of any planet; and even that, if known, can but little affect any conclusion as to the existence of inhabitants, since even with our own organised frames we know what great differences of temperature can be sustained without inconvenience; even enormous differences have been endured without actual injury; and the conception of modified organisations to suit any difference of temperature is by no means a difficult one.

The moon. Of the moon, from its proximity, we of course know more than of the rest of our system. The visible details of its physical conformation—or, as a recent writer somewhat strangely called it—the *geology* of the moon, are familiarly known, and lead to the general idea of a vast waste of extinct volcanoes without traces of seas or rivers, without a perceptible atmosphere, unless it be one of the most contracted kind, not reaching so high as to the tops of the mountains.

The most delicate thermoscopes have failed to detect the slightest heat in the concentrated rays of the moon; that is, in the sun's light scattered and dispersed away as it is after reflection from her convex and rough surface.

The intense heat which the moon herself acquires by absorption of these rays, if it could be radiated in sensible intensity to so great a distance as to the earth, would probably be all absorbed by our atmosphere and clouds, if not by any other cosmical medium, before it reaches us.

Yet conjectures are not wanting on the other side; according to which the absence of water has been accounted for by its transfer from the surface, for a whole fortnight exposed to the sun, to the opposite side "by distillation *in vacuo,* after the manner of the cryophorus."* These and other arguments must at least leave the question fairly open to discussion. Again, with regard to the visible hemisphere of our satellite, it has been remarked, that if the lunarians

* Herschel, Outlines of Astronomy, § 431.

are beings of at all similar social nature with ourselves, they would naturally inhabit cities; and large cities would be fully visible to powerful telescopes. Again, if they had extensively the use of fire, and manufactures dependent on its use, this would hardly fail to reveal itself by smoke. From the absence of such indications, then, it is affirmed that they have not anything resembling our manufacturing towns.

But against this it has been very recently contended,* on the strength of analogous actual observations, that at such a distance cities and smoke would not reflect light so as to be distinguishable.

Lunarians necessarily very different from terrestrials.

At any rate we see enough to assure us that the inhabitants, if any, must be very differently constituted from ourselves, to live in enjoyment under the manifest singularities of climate and physical condition and the extreme alternations of temperature to which they must be subject. But geology is not without its use in hinting at some kind of analogy in the moon with stages of formation which our earth has passed through. When the lunar volcanoes were active, the elements of water must

* See a Paper by Prof. C. P. Smyth in the Edinb. Phil. Journal, 1855.

have been present; and the terrestrial water, and even the atmosphere, must have been in a very invisible state when the earth was at an extremely high temperature. It may be added, that Mädler and Beer, after their accurate and elaborate examination of the physical structure of the moon, conclude with a distinct speculation on the probability of its containing inhabitants, however differently constituted from ourselves.

But all the foregoing speculations assume that the hemisphere of the moon, which is towards us, is similarly conditioned to the rest of its surface. Now, even at a very recent date, some results have been obtained of a highly curious nature with respect to the moon, which materially affect the question of its being inhabited, and show a peculiarity in the *visible* hemisphere.

Professor Hansen* has lately pursued some elaborate computations of certain corrections necessary to be applied to the inequalities of the moon's motion, deduced on the theory of gravitation, and thus involving as an element the attractive force of the

* Astronomical Society's Notices, vol. xv. p. 14.

moon, dependent not only on its absolute mass, but also on its figure. Hence he arrives at the theoretical conclusion that it is necessary to suppose *the centre of gravity of the moon different from the centre of its figure, and farther from the earth than the latter;* in other words, the hemisphere of the moon turned towards us is *more* raised above the *mean level,* as referred to the centre of gravity, than that away from us, to which its ocean and atmosphere, if it had any, would conform: this elevation at the middle point of the hemisphere amounts to nearly twenty-nine miles. And the summit of a mountain or table land of that height would be necessarily destitute of water or atmosphere; though both may exist with all the attendant phenomena of life in the hemisphere away from us, which is not thus raised; and even at the boundary some traces of an atmosphere might be perceptible.

Mercury, Venus, Mars.

The intensity of the solar rays, which would occasion so hot a climate in Mercury, may be greatly mitigated by the density of the medium by which he is surrounded; and the temperature to which the inhabitants of any part of Venus may be exposed will necessarily be very variable, from the extreme inclination of her axis, which gives rise to two winters and two summers in the course of her short year to the greater part of her surface, as is the case on a less notable scale within our own tropics, her tropics being near her poles.*

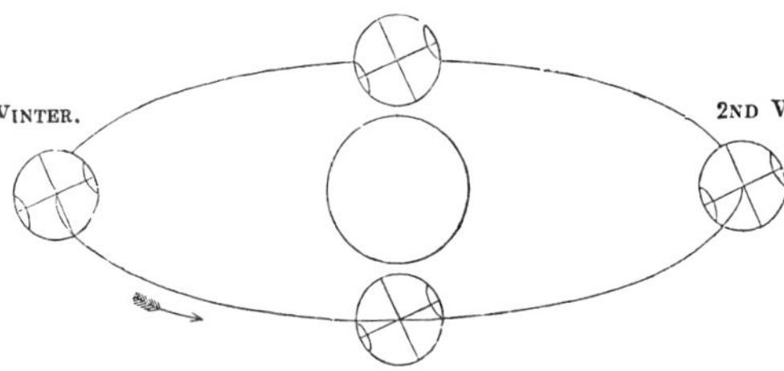

SEASONS IN VENUS. Inclin. 75°.

The planet Mars is on all hands admitted to be circumstanced so similarly to our earth, that little discussion can be needed.

Planetoids.

The kind of creatures capable of inhabiting the asteroids, or planetoids as they are now more properly called, is a question on which Sir J. Herschel † has not thought it unfit to offer some conjectures.

---

* On this point (as, indeed, on most others of interest connected with the physical facts of the planetary worlds), the reader will perhaps hardly need to be formally referred to Admiral Smyth's "Cycle of Celestial Objects," vol. i. p. 107., &c.

† Outlines, § 525.

But if they are uninhabitable — if they are exceptional in this respect — so they are in all others. Their number, so far from being exhausted by modern discoveries, will very probably be found to be infinite, and of all sizes, from the first known, downwards; not fragments of an explosion (which savours of the convulsionary hypothesis), but globules or molecules, gradually condensed *from a remaining ring of cosmical matter*, other parts of which *may* even yet be destined to undergo further condensations and combinations, and to form larger planets, which may not for myriads of ages be approaching the stage of bearing inhabitants.

Jupiter and the superior planets.

The materials of which Jupiter is composed are of a specific gravity about equal to that of water, which is the same nearly as that of the sun. The essayist, in his assumed magisterial vein, lays it down as by no means an arbitrary hypothesis, that Jupiter IS a globe of water; and argues accordingly that his inhabitants, if any, must be aquarian creatures of a soft, pulpy, boneless, watery character, to which, he thinks, we should naturally feel it very difficult to ascribe intelligence or moral attributes, that is without violating those analogies which we are so prone to form (perhaps groundlessly) from contemplating our own species.

But, as Sir D. Brewster, on the other hand, very justly observes, there are many solid substances, and even some minerals, as pumice, pitchstone, &c., and the metals of the alkalis, of less specific gravity than water. Jupiter, therefore, may just as well be composed of solid materials, and be tenanted by animals capable of living on land, as by aquatics. Similar calculations have shown, that in all the outer planets the conditions of gravitation are nearly the same; nor need the small specific gravity requisite for such animated beings occasion any difficulty. On our own planet, animals differ widely in this respect. It is hardly necessary to remark that birds, *e.g.*, have their bones, coverings, &c., of much greater specific lightness than the corresponding parts of terrestrial animals.

Nay, Sir D. Brewster has shown, by direct calculation, that even a human being, constituted as we are, would not really be much inconvenienced if transported to the surface of Jupiter; and buildings and trees, such as occur on our earth, might grow and stand secure, in so far as the force of gravity is concerned; and the same would be true for the planets exterior to him.* At any rate, when we reflect on the extremely varied forms of animated life on our own globe, on the diversified structures of different classes of animals, and the marvellous adaptations of their respiratory and circulatory functions to the conditions of their existence under the most varied circumstances, yet all preserving the most recondite relations to analogy and unity of composition, we conceive there can exist no difficulty in *imagining the possibility* of living beings constructed with bodies of greater or less specific gravity, suited to the most widely different conditions of gravitation or atmospheric pressure in which they might be destined to live, and with respiratory, muscular, digestive, or locomotive powers and capacities developed in infinitely varied degrees, according to the different conditions under which they might subsist, and the media in which they might have to move — yet always preserving an unbroken analogy with *some* grand and universal scheme of uniformity, of which we enjoy only partial glimpses; while under any

* Brewster, p. 64.

such variety of external form or condition, they *may* be equally capable with ourselves of being the recipients of higher principles of intellectual, moral, or spiritual life.*

Argument from the reflected light of the planets.

One of the arguments respecting the physical condition of the heavenly bodies, has been ingeniously derived from the polarised condition of the light reflected from their surfaces. In our own globe, this modification exists in the blue light of the sky. Sir D. Brewster, in referring to this fact, alludes to conjectures which assign, on this ground, an *aqueous* nature to the medium reflecting that blue light; and assuming that the fact is so, he goes on to apply the remark by observing that, to a distant spectator, the polarised condition of the light reflected from the earth would be a proof of the *aqueous* nature of its envelope, and hence such polarisation, observed as it has been in the light of the planets, is a like proof of the existence of *water* in them.† This inference, however, appears to me unfounded: all

* While this Essay has been in the press I find other publications on the subject announced; especially one entitled "Scientific Certainties of Planetary Life," by T. C. Simon. London, 1855.

† Brewster, p. 54.

transparent bodies, not only water, polarise light by reflection: and even opaque and rough bodies, which scatter the light irregularly, and even metals (contrary to the assertion of many elementary books) all polarise some portion to a very sensible amount. Hence the inference of an *aqueous* reflector is not a necessary one. In point of fact, the light of the moon and of comets has been found to be polarised, where the presence of water is more than doubtful. If the planets presented plane surfaces, or if we could otherwise determine the *angle* of polarisation, then the inference as to water might be verified.

Comets. The condition of comets is perhaps in some respects as well understood as that of the solid planets, notwithstanding their apparently more singular appearance, and, in some instances, enormously distant wanderings. They are certainly *transparent as masses*, whether composed of gaseous matter, or much more probably of minute solid molecules loosely aggregated, and probably kept so by a high state of electric tension, through whose interstices light passes, but which may also be transparent themselves. It may be too hasty an assumption to assert that even in their present state they are not likely to have organised inhabitants; yet they may at some future period be destined to *condense*, as Gambart's comet, separating into two, and each exhibiting a nucleus, shows a tendency to do; and may in time become fixed and consolidated bodies of our system, the future abodes of life.

Planetary comets and asteroids,

With respect to that singular system of comets of short period, consisting of at least five or six members, whose aphelia (or greatest distances) lie all within a little beyond the orbit of Jupiter, and which are all mere vast nebulous masses, yet as much parts of our system as any of the solid planets, it has been distinctly shown by Leverrier that they have probably been once traversing the depths of space in infinitely long orbits, and have at length been fixed in our system by the attraction of Jupiter. May they not, in the revolution of ages, possibly be destined to become more solidified members of our world? Nor can an astronomical imagination divest itself of the idea of some possible relation between this system and the ring of planetoids or scattered planetary globules, approaching in so many respects a cometary nature, which occupies the region at about the mean distance of the planeto-cometary bodies. In all these we may at least be at liberty to fancy stages of progress towards worlds, and that consequently in each there may exist at least the germs and seeds of organisation and life.

From what was before remarked as to the nature of the solar rays, arises this remarkable consideration with respect to comets, at variance with what is commonly supposed, that being, as we know they are, extremely *transparent*, however near they may approach the sun, his rays will *pass through them without heating them.* At least it can only be from *imperfect transparency*, which, doubtless, may be increased as they are more *condensed* at perihelion *, that any solar heat can affect them. But even then it may be much less than commonly imagined. This consideration may not be without value in reference to the conceivable idea of minute organised beings, monads or animalculæ, peopling the fine molecules of which cometary matter may consist.

In general, so far as anything is made to appear

* See Professor C. P. Smyth on Comets, Edin. Trans. 1850.

on certain and definite astronomical grounds, the whole question appears to be one of *degree:* the conditions of gravitation and atmospheric pressure, of heat and light, and the like, are only questions of *degree* in reference to all the planets of our system: they all resemble the earth *more or less,* they enjoy *degrees* of these *same* physical influences *more* or *less.* But their conditions in *no* instance differ *in kind.* From the brightest, hottest, or most intensely gravitating, up to the coldest, darkest, and most feebly attractive, there is but an enlarged or contracted *scale* of influences, and not a change in their kind or *nature.* So far as these considerations go, we can therefore see no proof or even presumption whatever against their possessing inhabitants; in structure and function, differing perhaps widely from those of the earth—in the adaptation and development of their forms or faculties modified in *degree* to as great an amount as the conditions of their existence may be diversified; —yet without any such *essential* and *radical* diversity in *kind* or *nature* as would place them out of all analogy with what we know of at least one small portion or member of the *great whole, or kindred scheme of cosmos.*

Conditions of planets differ only in degree from those of the earth.

Analogy from past state of the earth.

The considerations furnished to us by geology from its disclosures of the past history of the formation of the earth's crust with its organised inhabitants are perhaps among the most material aids to any inferences as to the structure and destination of other planetary bodies.

Earth long without man.

And here one of the points most dwelt on is the assertion of the recent date of man on the globe, which has been commonly assumed to be settled on what is confessedly mere negative evidence. It is, indeed, an opinion at present current among geologists, that man cannot claim a higher date than a period *later* than the latest of the tertiary deposits; but how many millions of years ago was the latest of these deposits is not so easily settled. This opinion is grounded solely on the *mere absence* of any remains hitherto detected, and with no powerful analogies in support of the negation, but with every probability to the contrary afforded in the apparent fitness of the state of the earth for man being its tenant at a much earlier date than that commonly assigned to his origin. There seems, however, to be a peculiar fondness in some minds for triumphing in the assumed recent origin of the race, not very intelligible on *philosophic* grounds, and in which both writers in the present controversy seem more or less to partake.

It is, indeed, on all hands admitted that we neither have any evidence, nor would analogy allow us to expect any, of the existence of man (or, indeed, even of the higher mammalia) throughout all the vast periods of the earlier formations.

But, after all, what does the argument amount to? Only to this: that the earth, during all these incalculably long past periods, was nevertheless in a state of *preparation* for man's habitation; and thus, although the planets may now be destitute of inhabitants, they may possibly be only undergoing similar changes anticipatory to a similar result; and thus, if the uninhabited state of the planets be admitted, we must still enlarge our ideas; we must embrace in our view not merely the *present* moment of time, but the whole expanse of the past and the future, and thus regard all worlds as equally *related*, actually or prospectively, to the destinies of moral and intelligent beings.

Planets may be for a time devoid of inhabitants.

Thus such considerations as geology furnishes are not without their use in pointing to a very possible interpretation of many parts of the planetary economy.

Yet afterwards become inhabited.

As our earth has undergone successive changes, and was probably for vastly long periods destitute of all organised life, or at least of all its higher forms; so it may be fairly argued, if any of the planets present features incompatible with organised life at the *present time*, it by no means follows that they will not have life developed in them at *some* future time, even up to its highest grade.

Or, again, others may possibly argue, they may have already attained that point, and have been since undergoing a retrograde influence of a destructive kind, reducing them to an uninhabited state. Yet surely, when all that we *know* of the past history of the universe, little as it is, is wholly stamped with the character of *advance*, we cannot easily reconcile ourselves to the idea of *retrogradation* or destructive agencies in any part or member of the system. But even were this so, it might be more philosophically regarded as merely one phase out of a vast series of changes in their recondite arrangement, to be succeeded again by renovated light and life.

In all that is urged relative to the physical conditions of the planets, there is nothing to show, that they may not experience changes in those conditions, or may not have already experienced them, in as great a degree as the earth, we know, has done, and beyond a doubt will continue to do: and this by the slow operation of immutably ordered, and profoundly adjusted series of regular laws and physical causes. If the earth in the process of progressive evolution has at a certain stage become the scene and site of intellectual and moral existence, the other planets also may have been so, or may at a future time become so.

Probable general diffusion of life, present or future.

Looking at the subject solely as a question of plausible philosophic conjecture, and guided as we should be by the pure light of inductive analogy, all astronomical presumption, taking the truths of geology into account, seems to be in favour of progressive order, advancing from the inorganic to the organic, and from the insensible up to the intellectual and moral in all parts of the material world alike, though not necessarily in all at the same time or with the same rapidity; in some worlds one stage being reached, while in others only a comparatively small advance may have been made.

Views of Œrsted.

It is on an enlarged view of the uniformity pervading the entire system of the natural world, both in space and time, that the assertion of rational inhabitants in other worlds has been defended by Œrsted in that eloquent posthumous work already referred to.* And though he agrees so far with the author of the Essay as to admit the possible *existing* absence of inhabitants in many worlds, on the same principle as in the case of the earth for such innumerable ages prior to the introduction of man, yet it is only on the truly just and philosophic ground of supposing universal and perpetual change and progress according to some grand laws as yet unknown, of which the existing condition of things forms but one small subordinate stage and portion, and which everywhere tends to prepare the way for the higher development of rational and spiritual life.

Œrsted has in fact pursued this topic in more than one part of his various essays, and on several grounds. In a portion of these reasonings he adopts a tone of somewhat vague metaphysical speculation

* The Soul in Nature, &c.

in which I cannot altogether follow him.* But so far as the present discussion is concerned, and in a more strictly physical point of view, the following passage will present a fair idea of the nature of his reasoning:—

"If we are now thoroughly convinced that everything in material existence is produced from similar particles of matter, and by the same forces, and in obedience to the same laws, we must also allow that the planets have been formed according to the same laws as our earth. This we know, however, that they have developed themselves during immeasurable periods of time in a series of transformations, which has also influenced the vegetable and animal creation of those periods. This development began with the lower forms, and advanced by gradual steps to the higher, till at length in the most recent periods a creature was produced in which self-conscious knowledge was revealed. We must, therefore, allow a similar mode of development in the other planets. There may be many which have not yet attained such a degree of development as our globe, or again

* See especially, pp. 53, 54.

other far higher beings may have been created: but everywhere the creatures endowed with reason are the productions of nature in the same sense as ourselves, that is, their understanding is bound up with the organs of their body; therefore the nature of their understanding cannot be fundamentally different from our own, but must obey the same laws."*

Admitting the general justice of this view, I must yet observe that this last idea seems to me to savour of that theory which would associate too closely the intellectual and spiritual nature of man with the physical, on the essential distinction between which I have before enlarged.

I will conclude with one more extract, which seems to complete Œrsted's view of the possible characters of the inhabitants of other worlds, and submit it without further comment to the judgment of the reader: —

"The variety in the nature of the planets of our own system is very great; but if we extend our thoughts over the whole universe, the differences are end-

* Ibid. p. 108.

less. On some planets the creatures may be possibly on a far larger scale, on others far smaller than on our own; on some, perhaps, they are formed of less solid matter, or may, indeed, approach the transparency of ether; or, on others again, be formed of much denser matter. The rational creatures on some of the planets may be capable of receiving far quicker, more acute, and more distinct impressions than on the earth, and on others it may be quite the contrary. If we now turn to the mental forces and mental development, we cannot acknowledge less variety. We may imagine that there are reasonable beings with weaker faculties than our own; but if we properly appreciate our present distance from the aspirations of our reason, we feel compelled to acknowledge that an endless number of degrees of development may exist above the point we have reached." *

* Ibid. p. 129.

§ II. — THE ARGUMENT CONSIDERED IN A THEOLOGICAL POINT OF VIEW.

Mutato nomine. — HOR.

The discussion has extended beyond the physical question.

THUS far I have endeavoured to restrict the view of the question to its *proper physical* and *philosophical ground — the extension of inductive analogies by reasonable conjecture.* But the writers on either side have not confined themselves to this view; they have in several respects availed themselves of *other grounds* of argument, and have had a special aim at ulterior and higher conclusions.

Radical defect of both views.

In the reference to such higher views, and the manner in which the question of planetary worlds is treated in its application to them, there seems much of a very questionable nature in the speculations on either side. But, in my opinion, the *radical objection* which pervades the whole discussion lies in the primary fact, that the question *is removed* from its proper *basis* of inductive conjecture and philosophical probability, and placed altogether on the new and unphilosophical ground of conformity to theological belief. The investigations may thus seem to be carried on, *not with a view to* PHILOSOPHICAL TRUTH, *but to serve an ulterior purpose;* a procedure which stamps the whole inquiry on either side with a character alien from scientific independence or freedom of inquiry.

Singular representations adopted in the Essay.

In the first instance, we cannot but be struck by the somewhat singular tone in which the physical question of the structure and origin of the planetary system is spoken of by the author of the "Essay," under the license of his anonymous mask: — "The planets and stars are the lumps which have flown from the potter's wheel of the Great Maker, the shred coils of which in His working sprung from

His mighty lathe—the sparks which darted from His awful anvil when the solar system lay incandescent thereon — the curls of vapour which rose from the great cauldron of creation when its elements were separated."* Again, the ideas broached with respect to the arrangements of the planetary system, and especially as affecting the rank and position of the earth, are characterised by a similar tone of paradox. Thus the author observes—"The orbit of the earth is the temperate zone of the solar system, and in that zone is the play of hot and cold, of moist and dry, possible."† . . . "The earth is the largest planetary body in the solar system . . . the vast globes of Jupiter, Saturn, Uranus, and Neptune, which roll far above her, are still only huge masses of cloud and vapour, water and air. . . . . The earth is really the domestic hearth of this solar system, adjusted between the hot and fiery haze on one side, the cold and watery vapour on the other. This region only is fit to be a domestic hearth, a seat of habitation; and in this region is placed the largest solid globe of our system; and on this globe, by

* Essay, p. 243.

† Ib. p. 196.

a series of creative operations entirely different from any of those which separate the solid from the vaporous, the cold from the hot, the moist from the dry, have been established in succession plants, animals, and man."*

Undue pre-eminence given to the earth.

Such representations seem somewhat to resemble the geography of the Chinese, where the Celestial Empire, stretched to gigantic proportions, occupies all the most central and fairest portion of the map, and the wretched inhabitants of Europe and America are condemned to its insignificant outskirts and remote corners.

But in the ideas thus broached it is not *mere* novelty and paradox on which I would animadvert. Such speculations appear like a *retrograde* movement. The author's masquerade assumes rather the fashionable *mediæval* costume. His theories carry us back to the schemes in old books with a gigantic earth, and a pigmy sun and planets performing their humble circuits round it. His speculations evince, if not a literal and physical, yet a moral *Ptolemaism:* they seem conceived in the

* Essay, p. 203.

spirit of the dark ages, which made man and his interests the sole centre and end of the universe, and would bind everything else in subserviency to them. The mediæval astronomy, as it made the earth the centre of the universe, so it helped to foster man's proud conceit, that he was himself the centre and sovereign of the moral creation. And this was intimately wound up with the superstitious belief of the age. The heavens revolved round the earth, to which everything was subservient; so man was the "microcosm," or emblem and type of the greater world without; and with him and his fortunes all nature was connected; to his benefit, or for his retribution, all things were made to be conducive. The planets shed their friendly or hostile influence at his birth. Their combinations and conjunctions were arranged to foretell his fate; and the return of comets and recurrence of eclipses were merely notices put forth to advertise nations and sovereigns of their approaching destiny.

The mediæval astronomy connected with theological views.

So, again, the theology of the age took up the same ideas, and made them the strongholds of its dominion; and the supremacy of an exclusive creed and the paramount dignity of the church were deeply concerned in the maintenance of the central position and immobility of the earth. As that position represented the metropolitan throne of the spiritual universe, which could in no way be consistent with a location in a subordinate and revolving planet, an insignificant globe among innumerable others—some far superior; so the earth's immobility typified the infallibility from which the authority of the church emanated, and whose pretensions could not but be placed in jeopardy when the former was assailed.

It should not be forgotten that the assertion of a plurality of worlds was one of the heresies for which Giordano Bruno was roasted alive; so that it behoves Protestants to watch the inroads of any theological dictation of the opposite view, which the tendency of the essayist's language seems to threaten.

Similar spirit in some of these speculations.

But though the precise conceptions and pretensions of infallible authority may have now disappeared, especially in Protestant countries, yet prepossessions of a very kindred nature, and evincing exactly the same *animus* and spirit,—"mutato nomine,"—prevail among some who profess themselves philosophers; and as they have been extensively evinced in regard to geology and other branches, so they have now influenced the discussion as to intelligent and moral inhabitants in the planets, as a question of *religious* import; and have led to the denial of their existence as derogatory to man's exclusive spiritual privileges and moral supremacy.

Inductive view.

But the true spirit of the inductive philosophy tends to teach a modest and just estimate of man's place on one of the smaller planets in a subordinate solar system of a subordinate stellar cluster, his whole race occupying but a speck in space, and *as yet* a speck in time: while it points by analogy to other similar worlds, possibly in different phases of development, whether corresponding to *future* stages of the earth's progress, or to *past*.

Relation of man to the system of the world.

I have elsewhere* endeavoured to illustrate and maintain the simple proposition, that *whatever is* animal in man's nature must be viewed as part of the same physical development and system, as the rest of animated nature: *Whatever is* superior to this belongs to a different order of conceptions, and cannot be affected by any physical considerations. On this view there is an obvious inconsistency in the

* See Essay I. § II.

desire to connect ideas of the spiritual nature of man with the laws of the material world, or to imagine the belief founded on them endangered unless it claim so uncongenial an alliance.

Man in his animal nature part of a series.

Equally groundless is the anxiety sometimes evinced to disconnect, as far as possible, even the *physical* history of man from that of the rest of the material world — to find breaks in the continuity of the order of nature — to represent what is called "the human epoch" in the world's history as marked and separated by some great gap or interval from all preceding epochs — to isolate man among animated beings on the earth, and to isolate that earth itself as his abode, among other worlds. The dogmatic assertion that "there is no * transition from men to animals" is clearly untrue in regard to the physical nature they possess in common; and, in this respect, we need not go even to the lowest form of savage life to find but too close an approximation to the brute.

Origin of civilisation

In the same spirit of viewing different parts of nature as *disconnected* from each other, and from

* See Essay on the Plurality, &c. pp. 81. 88. 164.

more comprehensive laws, much speculation has arisen on the history of civilisation, but on very insufficient data, because we *know* absolutely nothing of its earliest epoch, or of the element of *time* so essential to its satisfactory investigation. Much objection is felt to supposing man's condition *progressive*, or linked in one chain from the lowest savage life up to the highest civilisation and advancement; but it is difficult to see on what philosophical or rational grounds. It is argued, inferior animals possess a capacity for improvement, but only to a very limited extent, different in different species. It is very probably the case that similar *limits* exist in each separate *race* of men: though from one race to another there may be an indefinite capability of progress: but, on such subjects, our limited historical experience does not yet furnish us with sufficient data on which to prove any theory. Within the historic period, civilisation advances only by the slowest and most insensible gradations: and is communicated from one race to another. We have no right to assume that its advance was ever more rapid, but probably slower nearer its origin. The difficulty of conceiving the transition from absolute brutality to high civilisation—from sounds little better than the inarticulate language of beasts, to artificial combinations, the index of mind and abstraction—arises mainly from the utterly insufficient ideas commonly admitted in accordance with the received chronology, as to the length of time necessary for such advances.

But all questions of civilisation, language, and general advance, resolve themselves into that of the development of *mind;* and this, *in so far as* it is physical, comes, of course, under the law of continuity with other physical phenomena; *in so far as* it is metaphysical, it belongs to a higher order of things, independent of physical laws, though governed doubtless by equally regular laws of its own; and if this be true with regard to the *intellectual,* in a still higher sense does it apply to the moral nature of man; and thus civilisation does not involve any breach of *physical* continuity.

Higher principle in man distinct from all physical considerations.

If it be affirmed that "man differs in his kind, and even in his order, from all other creatures," it is certainly not in his material nature or animal instincts, but only in a higher sense. If it be asserted that the introduction of reason and intelligence upon the earth "is no part nor consequence of the series of animal forms; it is a fact of an entirely new kind: the transition from brute to man does not come within the analogy of the transition from brute to brute;" — to this I can only assent if understood as referring to that higher nature which is not amenable to physiological or material theories.

If in man's intellectual and moral development there is a new principle superadded, this is a metaphysical conception independent of any physical conditions. But if such a principle be superadded in one race of beings or in one world, there can be no physical reason why it should not be so in other races and in other worlds.

Illustration of this distinction.

To give a more precise illustration:—IF it were physiologically true that there were any peculiarity in man's *organs of utterance*, enabling him to frame articulate sounds, which is wanting in apes, then the cause of his superiority in this respect would clearly come under the dominion of *physical law*, and would mark a place and grade in *the connected scale of animal organisation*.

Or, again, IF any peculiarity could be shown in

man's *brain* to confer powers of *abstraction, moral consciousness*, or the like, which is deficient in the animal brain, this in like manner would indicate a clear physiological distinction, and would bring the case under the category of *degree of physical organisation, or development.*

But, on the other hand, if no such anatomical distinctions exist, then the source of the difference would be, as clearly, one *beyond the range of physical science* or material analogies. And thus *either way*, we must fully recognise the law of *continuity* as connecting man with the rest of the animated world: in the one case, because the transition would be simply one in *physiological* character; in the other, because there would be no break of a *physical* nature at all.

The relation of the animal man to the intellectual, moral, and spiritual man, resembles that of a crystal slumbering in its native quarry, to the same crystal mounted in the polarising apparatus of the philosopher. The difference is not in physical nature, but in investing that nature with a new and higher application. Its continuity with the material world remains the same, but a new relation is developed in it, and it claims kindred with ethereal matter and with celestial light.

Condition of the earth unconnected with spiritual nature of man.

Again,—the essayist would make the earth, in fact, the boundary planet between those too near the sun, and those too remote, to be capable of intellectual or moral life on their surfaces. He would represent it as alone of the proper density for mind to grow upon—as enjoying the precise proportion of light and heat for moral feeling to ripen, and the exact degree of atmospheric pressure under which spiritual aspiration can ascend!

Now on the *material* hypothesis there might obviously be a reason and a consistency in insisting on these differences between a hotter or colder, a moister or a drier planet, regarded as the parent soil of mind and spirit. But this connexion is entirely wanting when that hypothesis is so strenuously denied, and the essential discontinuity and absence of all relation and dependence between the development of man and the physical evolution of the material world so strongly asserted.

If the highest aspirations of man, the relations of his spiritual existence, be of a kind wholly independent of all physical evolution, and the very conception of them derived from teaching of quite another kind than any physical philosophy can supply, it would then be a question wholly alien and irrelevant, whether the earth were hot or cold, moist or dry, solid or aërial; the greatest or the least of planets—the sole inhabited world, or the most insignificant among the myriads of a peopled universe.

The question agitated in the publications before us has not been without its partisans among continental writers; though it must be said in general that science on the Continent is happily kept far more strictly on its own ground, and free from theological bias than among ourselves.

Œrsted (in the work before cited) alludes to some writers who, as he observes, "from one-sided religious or poetical views" have of late denied the existence of rational beings in other parts of the universe, in order to exalt the exclusive dignity of man. Œrsted's views.

He glances at the different races of inferior beings who in past epochs have tenanted the earth, and infers by parity of reason that other races superior to man in his present condition, may at future periods in like manner arise. He then proceeds to argue very much on the same general kind of mixed metaphysical and religious ground as his opponents appear to have taken, that

"Our entire system* has developed itself in a series of natural periods similar to the earth, and that each planet must still submit to a succession of creative transformations: consequently, we may infer that they have all had a succession of created beings, with such variations only as the different natural conditions of each must induce. Would it not be a strange assertion that neither the older planets at the most remote distance from the sun, nor the younger and nearer ones, had any of them attained to such a degree of development as is exhibited on our earth? Though a slight colour of support might be given to the assertion, it could never bear a close investigation. . . . Our system is but a small part of a far higher system, with which it has been developed under similar laws. . . . And must we believe that on none of these planets, similar or dissimilar to our own globe, reason has

* Soul in Nature, p. 53.

been awakened to self-consciousness? . . . . Thought never finds repose, but rises to higher and higher worlds; and except on earth, can it recognise nothing but barren solitude where no reasoning being has ever penetrated? No, it belongs rather to the nature of things that reason should develope itself into self-consciousness, not only in one spot, but in every member of the system, although in different degrees. . . . . . If we regard the whole of existence as a living revelation of Reason in time and space, we can conceive that the most varied degrees of development may be found distributed through all time; and that some bodies are still spheres of vapour, others have reached fluidity, while others have gained a solid nucleus, and so onward to the highest point of development, and then backwards again even to those bodies which are on the verge of final destruction. But even were it possible to maintain that self-conscious reason alone existed on earth, it still remains true, and is proved by the remains that have reached us of an earlier stage of development, that there was an immeasurably long period of time before the creation of man. Is it possible, then, that during the whole of this long period there was not a single being capable of perceiving and apprehending his own existence?"*

I have given this passage at length, in order that it may be seen how far speculations of a kind necessarily somewhat vague may influence a mind of such philosophic capacity as that of Œrsted. At the same time, without professing to admit the entire force of such reasoning, I am disposed to concur in the general conclusion as at least a far more consistent and worthy belief than that which would narrow and restrict all intellectual and moral existence to the confined limits of our little planet.

Final causes appealed to in this argument.

But apart from these speculations, other considerations of a more distinctly theological kind have formed throughout the acknowledged basis of the reasonings of both disputants; and, indeed, the main motive for pursuing the inquiry. In the first place, then, the whole discussion has been closely connected with the *argument from final causes.*

As this mode of argument is avowedly and extensively adopted by one writer, and though much restricted and qualified, is yet in some sense referred

* Soul in Nature, p. 54.

to by the other, it will be desirable to recur briefly to the general grounds on which such a line of argument can be sustained; but as this is a topic which has been already dwelt upon in a former Essay*, it will not be necessary here to do more than refer to what has been there said, and to proceed upon those grounds to apply the argument to our present subject.

Reference to Professor Owen's archetypal theory.

In the first place, the profound generalisations of Professor Owen have been referred to, who, in his discussion of the vertebrate skeleton and its theoretical archetype, dilates on the conclusion, that besides the organised structures *actually developed* on the plan indicated, according to the same principle the *rudiments* of an infinite variety of other such forms exist, and may therefore possibly remain to be developed; he observes that such conceivable forms are far from being exhausted in existing or past life on this globe; and that, "though they may never be developed as such in this planet, it is quite conceivable that certain of them may be so developed, *if* the vertebrate type should be that on which any of the inhabitants of other planets of our system are organised." †

* See Essay I. § v.

† On Limbs, p. 83.

Again, carrying out this idea to the structure of the eye and its possible modifications, as connected with the vertebral theory, the author argues that in Jupiter, with such provision for illumination as exists there, and with the same laws of light and other similar conditions, *analogy* would lead us to infer the *probability* of beings with eyes conformably modified; and such creatures, he says, "*may* exist to profit by such sources of light; and *must* exist IF the only conceivable purpose of those beneficent arrangements is to be fulfilled."

Now I quote these words more especially with a view to remark *the nature of the reasoning:* we cannot but observe the truly philosophical tone of caution, united with the legitimate pursuit of analogy, with which the distinguished author pursues his conjecture. The reference to the archetype is simply one of the highest forms of *conjecture* from *analogy*, and supplies the same kind of antecedent presumption, which the existence of a theoretical mathematical formula would do in guiding us to a physical truth.

That other modifications of the primeval type not carried out into actual being on our planet, may possibly be so in others, is abstractedly a very fair conjecture; and the existence of such unrealised cases here may no doubt afford something like a presumption that they may be realised in other planets. So far, then, the reasoning is simply and strictly of an inductive character.

But the further remark such beings "*must* exist," is, with equally just and philosophic caution, qualified by the condition, IF the purpose "*is to be* fulfilled." In a word, THE ARGUMENT FROM FINAL CAUSES is here kept properly distinct from that of INDUCTIVE ANALOGY, and is only maintained on the express *hypothesis* that we may reason at all from a purpose to be fulfilled.

Improper introduction of final causes in science.

On the general admissibility of such a reference to final causes, we must recur to the observations made in the former essay. If the principles there laid down are admitted to be just, we shall the more readily acknowledge the general impropriety of attempting to solve a philosophical problem like that before us on any other grounds than those of legitimate physical analogy.

The argument from final causes is, in one sense, wholly distinct from any of a purely philosophical or positive kind; in another, it may be understood as a simple extension or higher theoretical view of the argument from analogy.

Thus, for example: Kepler argued from final causes to his first conjectures of the laws of the planetary motions. "I reasoned," he says, "that if God had adapted motions to the orbits in some relation to the distances, it was probable that he had also arranged the distances themselves in relation to something else." But this was nothing more than a *guiding conjecture*, leading him to try *inductive* processes.

With more special reference to the question now before us, we find in numerous instances a purpose answered: we infer it *probable* that in others, or in all, it may be so likewise. In one case, we trace a structure adapted to a particular end; in another, under apparently analogous conditions, we infer that a similar end *may be* answered. In our earth there is a certain provision of light, and there are beings with eyes adapted to enjoy it. In Jupiter there is a certain, but different, provision of light: by analogy, there may be beings with eyes equally capable of enjoying it, though in a different degree.

The conjecture is perfectly fair and philosophic *in its nature;* but other concurrent circumstances must be taken into account before we can obtain any higher amount of probable evidence. Nevertheless, as far as it goes, as *suggestive of presumptive probability*, it is strictly legitimate.

But the subject has been carried out further by the introduction of a still more metaphysical kind of argument deduced from the "archetype" considered as a revival of the Platonic idea of such archetypes existing in the Divine mind.* On this point the authority of the erudite Cudworth has been appealed to, and the quotation of a passage from the "Intellectual System" has given rise to some discussion, which, however, seems to me to have little real bearing on the question. I merely observe that the argument thus derived from our belief in the Divine attributes or the assumption of intentions or ideas in the Divine mind, whatever may be thought of it in a metaphysical or theological sense, does not in my opinion belong to the province of physical philosophy; nor can it, I conceive, be

* See Brewster, p. 84. Supplement to Essay, p. 27.

legitimately introduced in any such discussion as the present.

Final causes urged in support of inhabited planets.

The argument from final causes is largely appealed to by Sir D. Brewster, and is closely connected with the peculiar religious turn he is disposed to give to the question of inhabited worlds. The belief that the planets and even the stars are inhabited, is upheld by him on the express ground of the probable end or purpose of their existence being no other than the support of organised life and intellectual and moral creatures: and that to reject such a belief is to involve the irreligious idea of denying the final cause of their creation.

Thus, it is argued, that large globes, attended by an apparatus of satellites, or rings, *must* have been created for some great and worthy *purpose;* and that *we cannot conceive* any such purpose but that of sustaining animal and intellectual life.* Again, as it is the obvious *function* of the sun to supply heat, so there is no *conceivable function* of the planets but that of supporting inhabitants.†

* Brewster, p. 84.

† Ib. 90.

The fixed stars "were not planted in space to shed their light and heat upon nothing."* "If the stars are not suns, for what conceivable purpose were they created?"† Again, if no life existed in the universe, he observes, the celestial movements would be going on "fulfilling no purpose that human reason can conceive—lamps lighting nothing, fires heating nothing, waters quenching nothing, breezes fanning nothing; and everything around, mountain and valley, hill and dale, earth and ocean, all *meaning nothing*."‡ Or, more pointedly, thus: "In peopling such worlds with life and intelligence, we *assign the cause of their existence*."§ "Life was not made for matter, but matter for life; and in whatever spot we see its atoms, whether at our feet, or in the planets, or in the remotest star, we may be sure life is there; life to enjoy the light and heat of God's bounty, to study His works, to recognise His glory, and to bless His name."‖

Objections to this reasoning.

Now, in looking at the application of this kind of argument in the present case, even if disposed to admit the *truth* of the conclusion, I should

* Brewster, p. 159. † Ib. p. 232. ‡ Ib. p. 181.
§ Ib. p. 179. ‖ Ib. p. 180.

still have much doubt as to the *mode* of arriving at it.

Such an argument, in the first instance, necessarily *presupposes* the fact that the conditions of the planetary bodies *are adapted* to be inhabited: the very point in question. But supposing the fact of such *adaptation* admitted, the next step in the argument is the assertion, "that *no other* end or purpose of the existence of planets can be conceived." Now, this must, on the slightest consideration, be allowed to be at least a very hazardous assumption. How can we undertake to affirm, amid all the possibilities of things of which we confessedly know so little, that a thousand ends and purposes may not be answered, because we can trace none, or even imagine none, which seem to our short-sighted faculties to be answered in these particular arrangements?

Supposing, however, that all this were conceded, it still remains to connect it with the conclusion because *no other end* can be assigned, therefore, *this one* end of sustaining life not only must be the *sole* real end which the Creator had in view, but must be actually accomplished in all the planetary worlds: an alternative and a conclusion which hardly appears warranted by any sound principles of reasoning.

The earth certainly was for myriads of ages destitute of human beings: it existed, therefore, in *vain for man.* Hence, it is an undeniable parallel, if the planets are now uninhabited, and therefore useless, so was the earth for an unlimited number of ages in past epochs. If the one be a contradiction to final causes, and to be rejected as inconsistent with the Divine beneficence, so must the other be: yet this other we know to be the fact.

On the other hand, it is truly satisfactory to find so able a writer as the essayist joining his own testimony to that of Professor Owen against the narrow view of final causes, and beginning to avow that more truly just and philosophical principle of openly confessing that we know not, and ought not to pretend to know, *why* this or that arrangement is made. The essayist has, in fact, elaborately argued this point in his eleventh chapter, and more briefly and boldly in one of his dialogues * observes,

Opposite views more philosophical.

* Supplement, p. 5.

"I do not pretend to know for what purpose the stars were made, any more than the flowers, or the crystalline gems, or other innumerable beautiful objects."

Again: "I have learnt much from Mr. Owen. I have learnt from him, in many most striking cases, to admire purpose in organic arrangements, when purpose is apparent. But I have learnt from him, also, that to infer facts from 'an only conceivable purpose' is a very hazardous process." *

Without any disparagement to Professor Owen, I conceive not merely this partial lesson, but a more extended one as to the whole ground of argument, might have been rather learnt in the school of Bacon, and the incongruity of narrower views with the essential spirit of induction; which would lead us rather to recognise uniformity of plan, *law, order, and unity, as the true exponents of design*, than to seek for mere utility and ends to be answered,—however important in a subordinate sense,—and to carry out such principles as our only safe guides in

* Supplement, p. 30.

speculations even in the region of imagination, whether exercised in peopling worlds or in depopulating them.

Yet unphilosophical conclusion on the other side.

Yet, notwithstanding these admitted considerations, the argument of the "Essay," in fact, rests much on final causes. It is with reference to *an end*, to the fond belief in the high importance of *man* in the universal scale of being, that so much stress has been laid on his *recent* date upon earth, and the alleged extraordinary peculiarity and singular isolation of his position, apart from all admixture with the animal creation; and this is more especially dwelt upon with a view to the argument that all the universe must be supposed a waste in order to enhance the moral dignity of one puny race, and to enable him to believe that his little world and his species are the peculiarly favoured objects of the Creator's care.

The earth was for myriads of ages a void, and for equally long periods tenanted only by inferior creatures, solely to the end that man might at length come in solemn pomp at the close of the long procession, and take possession of his throne! and not only so, but all the most distant planets and remotest worlds, invisible to his eye or to his telescope, are destined to a similar humiliating inferiority, solely to swell the triumph of his supremacy, and to exalt the dignity of that little speck on which the mighty displays of Divine power and mercy were to be made for his exclusive benefit!

Thus, then, we find, in point of fact, the argument from final causes applied with equal force to support diametrically opposite conclusions. Tacitly referred to on the one hand, it clearly evinces the uninhabited condition of all worlds but our own, because man alone is privileged to be the exclusive recipient of the Creator's beneficence; openly and strenuously upheld by the other disputant, it as manifestly shows that the planets, and even the members of the most remote sidereal system, must all be teeming with rational and spiritual beings to exalt the same Creator's perfections, and render a reason for their existence.

On the one hand, it is argued that the planets *must be inhabited*, because they could only have been created for the sustenance of life; on the other, that they *must be uninhabited*, because they could only have been created as foils to enhance the dignity of the earth and of man. On the one side, the universe must be inhabited because a void universe would be *useless;* on the other, a void universe is *necessary* for the exaltation of man and of the Divine dispensations towards him.

Bearing of the argument on natural theology.

If in a more wide and worthy sense we come to consider the religious application of the argument, whether for or against other inhabited worlds, it will be easily seen that, under any point of view, it amounts to little. With regard to the great truth of natural theology, the evidence which cosmical order affords for a Supreme Intelligence is in no way affected by the question of a plurality of worlds; it stands unassailable on the basis of demonstration; and can be little affected by any further speculative arguments. If the existence of inhabitants in the planets were as much demonstrated as on the earth, it would undoubtedly enhance the great argument by the extension of its evidence which would be furnished by the existence of organised structures, or of intellectual and spiritual beings, from one member of the system to many others: and if *universally* proved, it would tend to exalt this branch of the argument in a proportionate degree — by the infinite multiplication of such instances of physical, moral, and spiritual existence.

No real addition to the evidence; but wider contemplation opened.

But *demonstration* has never been in the slightest degree pretended to in this matter; the very utmost which its warmest supporters have claimed has never been more than *analogical probability.* Its force, then, as an argument of natural theology could weigh nothing in comparison with those substantial evidences which the *demonstrated* facts and laws of science afford. The utmost that can be said of such a theory is that, if admitted, it affords a beautiful opening for a more extended religious contemplation of the Divine beneficence reaching to so many more myriads of creatures capable of estimating it. Indeed, without unduly pressing the argument, we might fairly agree with Sir D. Brewster that, when we contemplate the combination of worlds upon worlds, and especially the movement of the entire solar system round the supposed *central sun*, "the mind rejects almost with indignation the ignoble sentiment that man is the only being that performs this immeasurable journey;" and that the planets with their train are but inert masses "mocking the creative Majesty of Heaven." *

Adopting the hypothesis of inhabited worlds, the devout believer in supreme and superintending Wisdom and Goodness, would doubtless find his adoration exalted and enlarged in proportion as he conceived a more enlarged sphere for its manifestation, and believed that myriads of other beings peopling other worlds, however different in nature from ourselves, were at the same time rejoicing in the light of the same beneficent Fountain of Good. Yet, he would recollect that all this is purely *hypothetical*, and stands on grounds quite distinct from the grand primary convictions of the unity and harmony pervading those worlds, and the consequent recognition of a Supreme Intelligence.

Œrsted's view of the subject.

I have before adverted to the views broached by Œrsted on the general question of inhabited worlds. Without professing to assent to all his opinions, it must yet, I think, be allowed that there is much force in some of his representations. He argues much on the *intellectual* capacities of the supposed

* Brewster, p. 123.

inhabitants of the planets; and more especially contends that *necessary* truth must be the same to them as to ourselves: though great differences might exist as to their perceptions of natural phenomena; and even in the former case, there may be great differences in the strength of the reasoning faculty.

To the same effect he continues a passage * before quoted, with respect to their moral development; and even goes further, and contends for a community of *moral* laws throughout the inhabited universe. Setting out from the theory of moral obligation on our own globe as arising necessarily out of the position and nature of man and his relation to his Creator, Œrsted argues that the same must hold good with the inhabitants of other worlds, making due allowance for the actual diversities in their conditions.

I will give one short extract:—

"Throughout the universe there are beings endowed with the faculty of understanding that they may be able to catch some sparks of the Divine light;

* Soul in Nature, p. 96.

and God reveals Himself to those beings through the surrounding universe, and rouses their slumbering reason by that Reason which reigns throughout the sensible world; nay, He gives them a deeper insight into material existence the more their own minds are awakened; and thus they find themselves placed in a ceaseless stirring development, which, after having reached a certain point, removes them further and further from the idea that the foundation of being is that which is palpable, and which leads them to acknowledge and view themselves, their spirits and bodies, as parts of one eternal organism of Reason."*

True unity of worlds.

Without professing to adopt or even to understand entirely the ideas thus eloquently expounded, I would yet willingly express an assent to the broad principle involved; and in such generalised conception (warranted as I think by the soundest inductive principles) it is that I would recognise THE TRUE UNITY OF WORLDS; not a narrow restricting of all development of mind and soul to one minute speck in the universe, which appears to me as unphilosophical as it is derogatory to the

* Soul in Nature, p. 109. See also p. 128.

worthiest conceptions of the Supreme Mind, but an enlarged admission of all worlds as harmonising parts in one great whole, and of the universality of such development of life, either "*in esse*" or "*in posse*," in actuality or in potentiality, as wide as is our belief in the universal presence and operation of the Great Source of all life, all mind, and all soul.

Including origin of worlds and past changes.

If we look back to past changes and the probable preparation for organised life even in planets not at present fitted for it; or if we ascend to the still grander question as to the probable order or law by which such beings may have come, or may yet come, into existence, though utterly unable to give any positive reply, yet we cannot fail to combine every reflection upon such a question with the great law of continuity, and beyond all doubt to regard as highly probable, some intimate connexion between the series of physical arrangements of unorganised matter, the successive gradations of organised existence, and the crowning of organisation with animal and intellectual life, making it the fit recipient for higher spiritual manifestations; and to recognise throughout the series the close dependence of the whole on some great principles of law and order (however unknown to us) to just the same extent as we acknowledge their dependence on supreme intelligence and power, exhibited to us through these its *cosmical* manifestations.

Secondary means the evidence of Divine operations.

The precise nature and order of those causes which brought about the evolution of organised life on our globe are *as yet* unknown to us, however open to plausible conjecture. But in proportion as they might be known, they would afford *increasing* evidence of supreme intelligence: *increasing*, just as a more complex self-adjusting machinery would afford higher proof of intelligence than that which wants manual regulation. I have observed in a former place that all rational natural theology proceeds by tracing the steps and processes in which design is evinced. The more steps in such processes we can trace, the more satisfactory our convictions; and if, where we do not *know*, we can fairly *conjecture*, the legitimate conjecture will have a like tendency.

Objections.

Sir D. Brewster *, however, considers it a highly objectionable idea, to suppose the planetary system

* More Worlds than One, &c. p. 249.

"*manufactured*" out of a nebulous mass by means of certain material laws. I, on the contrary, would accept this phrase, and contend that it is precisely the idea of their being *manufactured* instead of *made*, which would constitute the stronger proof of *intelligence* and *mind.*

Precisely in proportion as a fabric manufactured by machinery affords a higher proof of intellect than one produced by hand; so a world evolved by a long train of orderly disposed physical causes is a higher proof of Supreme intelligence than one in whose structure we can trace no indications of such progressive action. And in proportion as we might be able to follow out more and more details of that succession of causes, should we derive increasing evidence of the great truth.

Bearing of these views on revelation.

But the religious contemplations connected with this subject have assumed also more definite forms, and have involved difficulties at first probably little to be suspected on other grounds than those yet adverted to. It has not been merely with a reflection on the enlarged beneficence of the Creator, that religious men have been contented to regard the supposed existence of planetary beings; they have also viewed the question in its bearing on the belief in a Divine revelation and the mysteries of human redemption. In this point of view, then, we must proceed to look at it.

Religious difficulties felt.

In fact, the main object in view, in both the works under consideration, is an application of this theological nature, and to furnish replies, though in very different ways, to certain objections felt on religious grounds to the doctrine of a plurality of worlds.

It has been held that the belief in the existence of rational and moral beings, however unlike ourselves, in other planets or other systems, is a notion which, apart from its physical vastness and difficulty, involves the believer in religion, whether natural or revealed, in perplexities and objections of the most serious nature, such as, in fact (it is alleged), seem only capable of being relieved by the rejection either of religious faith or the idea of a plurality of worlds. These difficulties and objections are dwelt upon with great emphasis, and are stated at large by both writers.* We learn that they have pressed upon the

* Essay, § 2, 3. 4. More Worlds than One, ch. vii., &c.

serious convictions of many excellent persons, not only of high religious feeling, but even of cultivated and enlightened minds; and have thus engaged the attention of powerful champions of Christian truth in the endeavour to remove or mitigate them. In particular, the arguments of Dr. Chalmers are referred to in detail; and while they are highly commended by the author of the Essay, are yet deemed hardly sufficient without the further extension which he conceives his own speculations confer on them.

Alleged difficulties in some instances vague and unmeaning.

Thus, in a kind of ironical tone, the supposed advocate of a plurality of worlds is represented as putting forth his theory in a somewhat dogmatical manner; and, after making his various assertions as to the nature of the remotest planetary systems as the seats of animal and vegetable creations, and the habitations of rational and responsible beings, is made to add, "the only matter which perplexes us is, that we do not quite see how *to put our theology into its due place* and form in our system."* Thus far, however, we do not exactly make out what the precise source of this perplexity may be, or how it is

* Essay, p. 121.

that theology can want a local habitation in a planetary or sidereal system at all.

We read through pages of eloquent and devout declamation, abounding with varied expressions of sentiments of religious humility and pious reliance on the care of the Creator—of profound conviction of His beneficent superintendence over all His works, and especially over man, so important in his *moral* and *spiritual* relations, even though insignificant in a physical point of view, and as measured by the immensity of the *material* creation—all which cannot but meet with cordial and general concurrence; but still in which there seems little to lead to any possible collision between such sentiments and any astronomical or cosmical speculations.

Again, the same "religious difficulties" seem to have made an equally powerful impression on some minds, whose perplexities are considered by Sir D. Brewster. A large portion of his work is devoted to the consideration of them; and it appears also that similar objections had presented themselves even in past times, when the speculations of Fontenelle and others were broached, and were of a nature to give a handle to the sceptics and scoffers of the day in their attacks on religion.

Difficulty from the insignificance of man.

One of the first of these religious difficulties which is at all distinctly brought out, is the notion (which however incredible it may appear, from the testimony quoted, it would seem, has really haunted some pious minds), that amid the multiplicity of creation, if believed to be replete with boundless worlds of life and intelligence, the humble individual, man, on this small speck of earth, *might be overlooked;* might be too small for the eye even of *Omniscience* to discern — too insignificant even for the vigilance of *infinite* and *omnipresent* Providence to take care of!—or that the *omnipotent* Creator "*may have too much to do*"* to extend his concern to all!

Every real religious doubt or scruple undoubtedly has a claim to be treated with the most unfeigned respect; but it is difficult to repress an expression of astonishment when we find that such objections as these should have been seriously entertained or have been deemed deserving of elaborate refutation by

* See Brewster, p. 149.

Dr. Chalmers, by an appeal (however forcible in itself) to the revelations of the microscope disclosing worlds of animalcular life equally the objects of providential care; or, further, that Sir D. Brewster should think this answer unsatisfactory, and actually go through some amount of reasoning to supply what he thinks a better!

Difficulty as to other worlds partaking in spiritual privileges.

But another difficulty next presses upon us of deeper import; we are told that the assertion of moral and spiritual beings in other planets is full of danger to the belief in man's high privileges, the possession of a special Divine revelation, the communication of spiritual gifts, or the promise of immortality. That these are, in fact, *exclusively* his portion and inheritance, the security of which would be hazarded by imagining any other claimants on such dispensations of the Divine mercy.

Spirit of exclusiveness.

There are doubtless peculiar charms in the *exclusive* nature of privileges, which, in fact, often constitutes the main value of their possession. Thus, in the present instance, a feeling of complacency and comfort seems to be inspired by the belief which, with a non-peopled universe around him, man can securely entertain, that he is the sole favoured child of his Creator, and can console himself in looking round on the untenanted planets with the happy reflection, "all these vast worlds were made, if not indeed in a direct way for my use, yet still for gratifying me with the delightful reflection that my own race on my own happily constituted planet are alone permitted to enjoy the blessings of life and intelligence, which are denied to other inferior globes, or to aspire to any of those higher gifts of grace or glory which are our peculiar inheritance." We cannot fail to reflect how highly and peculiarly spiritual is such a contemplation, and how eminently worthy to be dilated on by a Christian divine! how powerfully tending to elevate at once the conception of a beneficent Creator and the moral dignity of the creature!

Moral training in other worlds.

Again, it is urged as a more specific objection, that, if we people the planets, we must by analogy suppose those inhabitants to have had a similar intellectual and moral progress and training to that which the human species have undergone. But this, it is alleged, would impugn the special character of God's government of our world, which consists only with "man's nature and place being unique, and incapable of repetition in the scheme of the universe." *

The "moral training" and the "religious education" of the human race and the like phrases, have become very common with writers of late years. I can only apprehend their meaning in the simple sense, that, as mankind have advanced in civilisation and intelligence, different codes of morality have been tolerated, and varied forms and dispensations of religion suited to those successive stages of advance have been established. But on what grounds it can be asserted that such a series of progressive movements are "incapable of repetition," if the circumstances of moral and spiritual beings should require it, whether on our earth or elsewhere, after all that has been urged, I am wholly unable to see.

To the same purport the author of the Essay again observes: —

"Unique position" of man.

"Religion seems, at first sight at least, to represent man's history and position as unique. Astronomy, some think, suggests the contrary: I examine

* Essay, pp. 36. 51.

the force of this latter suggestion, and it seems to me to amount to little or nothing." *

In what way religion "seems" thus to point to the unique position of man, is precisely the point which I fail to apprehend. I see nothing in natural religion to lead to this conclusion; and in the pages of revelation, even in its most literal acceptation, whatever privileges are conceded to man are surely *nowhere denied* to other *possible* races of beings.

Assertion of revelation only made to man.

But, more precisely, the representation takes this turn; man, it is contended, is the peculiarly favoured creature of Heaven—the earth in which he dwells the scene of the most peculiar and transcendent displays of Divine mercy: the human race has peculiar and exclusive relations with the Deity. The history of this race attests a continued scheme of exclusive dispensations of grace. And we are told "The earth, selected as the theatre of such a scheme of teaching and redemption, cannot in the eyes of any one who accepts the Christian faith, be regarded as being on a level with other domiciles. It is the stage of the great drama of God's mercy

* Supplement, p. 42.

and man's salvation; the sanctuary of the universe; the Holy Land of the creation; the royal abode, for a time at least, of the Eternal King." *

That such privileges are granted to man, that such displays of Divine mercy have been manifested, is no doubt clear to every believer; but why they must be supposed *exclusive;* why *limited* to man and his earth; why *denied to all other* possible or conceivable races of beings — I am at a loss to understand, or to find a particle of evidence in support of such an extraordinary assertion.

It is, professedly, to meet the fatal difficulty which is thus believed to result, if we deny the *exclusiveness* of human privileges, that the essayist has engaged in the arduous task of attempting to prove that there *cannot* exist in any of the celestial bodies inhabitants of a moral or religious nature to dispute with man his *exclusive* privileges. If this conclusion be made out, of course the objection, such as it may be, falls to the ground. But if the point of uninhabited worlds be not *demonstrated* (and who can pretend to say that it is?), then it would be

* Essay, p. 44.

implied that the whole faith of the Christian world is in jeopardy: and its vindication and truth are staked upon the acceptance of the belief in an uninhabited universe, which can be nothing more, at best, than a visionary speculation!

Constitutes no real objection.

But if we look further at the real nature of this difficulty, and endeavour to put it into more definite shape, I conceive it can only be stated somewhat in this way:—A special manifestation of the Deity in the Gospel dispensation is affirmed to have been vouchsafed to the inhabitants of the earth. But the earth is a very small and insignificant unit in a vast universe of similar and greater worlds, all teeming with unnumbered moral and spiritual beings possibly of far higher dignity than man. *Therefore*, we are to doubt the reality of the revelation to us!

Antecedent probability the same for other revelations.

But, perhaps, it may be said the objection only refers to the *general antecedent probability* of a revelation (in the accepted sense of the terms), and does not descend to the question of its particular alleged *evidences*. Now, as to this question, we have only to consider in what possible way the assertion or denial of inhabitants in other worlds can affect the antecedent probability of a revelation being made to the inhabitants of the earth. Taking the argument for probability as stated by the most approved writers on the "evidences," it can be maintained only on the *same general grounds* of the Divine attributes, which would render it *equally admissible* in regard to the supposed inhabitants of any other worlds—the creatures of the same Supreme Power, to whom, by parity of reason, it must be supposed a revelation *would* be equally granted by the same Supreme Goodness *if* they needed it, and which assuredly *could* be accomplished under whatever diversity of condition by the same Universal Omnipotence. On the other hand, if those grounds of argument, so commonly adopted, are themselves thought unsatisfactory in regard to the planets, why are they admissible with regard to the earth?

Analogy would extend revelation to other worlds equally.

If analogy did assign intelligent and spiritual inhabitants to any number of other worlds, or if it represented the whole universe as nothing but one teeming creation of moral and spiritual existence, surely the same analogy would not only admit, but rather require, the extension of the same argument for the probability of Divine communication with one portion or race of such beings as with another; or rather, it might even seem to suggest the notion of one grand universal manifestation of the Divinity in all the worlds over which the same universal Providence presides.

Objections from peculiar views of man's condition.

But the objection has taken a more specific form from a reference to certain deep doctrinal views relative to the moral and spiritual condition of man, and especially the sinfulness of his nature, which, it is alleged, are endangered by the consideration of other inhabited worlds. *Yet surely*, if those beings are in their spiritual needs similarly circumstanced to man, it is the fairest presumption that corresponding means of spiritual restoration would be granted them. If they do not require those means, they may equally stand in other relations adapted to their nature and condition. Yet this kind of objection seemed to Dr. Chalmers so formidable as to require elaborate refutation, and Sir D. Brewster also makes a lengthened reply; in the course of which he plunges into the depths of a metaphysical theology, the theory of original sin, and the hopeless question of the existence of evil; which last he seems disposed (unless I misunderstand him) to solve in the convenient way so readily adopted by some other philosophers and divines, of *denying its reality*, and affirming that "the spectre of moral evil has been conjured up by ourselves." *

Origin of much of the supposed difficulty in peculiar views of Christianity.

Without pretending to go into such discussions, which appear to me equally needless and interminable, or dwelling on the hazardous nature of this mode of reply, it will suffice to remark generally here that these topics seem chiefly connected with the present subject and with the question of the origin of the human race only from the prevalent adoption of the theology of a peculiar school, according to whose system certain supposed *physical* changes were induced by sin and "the fall;" to which in vulgar estimation such singular effects have been attributed. It ought surely to suffice a reasonable and Christian inquirer to refer to the language of the Apostle Paul †, and to perceive that such ideas can find little foundation even in the most literal acceptance of his words, which do not contain the smallest allusion to any *physical changes in man's nature*, but to a subjection to death in

* Brewster, p. 138.

† As *e.g.* Rom. v. 18, 19.

Adam *in the same sense* as accords with a deliverance from it in Christ.

More precise statement of the main difficulty.

In pursuing the argument, however, Sir D. Brewster puts the difficulty, at least more distinctly than before, into the mouth of a supposed timid Christian, "How can we believe that there are inhabitants in the planets when God had but one Son whom He could send to save them?" and adds, "If we can give a satisfactory answer to this question, it may destroy the objections of the infidel, while it relieves the Christian from his anxieties." To this task he therefore addresses himself, putting the case thus:—"When at the commencement of our era the Great Sacrifice was made at Jerusalem, it was by the crucifixion of a man, or an angel, or a God. If our faith be that of the Arian or the Socinian, the sceptical and the religious difficulty is at once removed; a man or an angel may be again provided as a ransom for the inhabitants of the planets. But if we believe with the Christian Church, that the Son of God was required for the expiation of sin, the difficulty presents itself in its most formidable shape." *

* Page 139.

Sir D. Brewster's answer.

The author's answer is to the effect that, "as by some process of mercy, which we understand not,".. the saving power of the sacrifice has been communicated alike to the most distant nations and ages, past and future; so it might just as easily be communicated to the inhabitants of the most distant planets and worlds; and if this should not be convincing, another answer is hinted at (though not entirely approved by the author), yet as satisfactory to some minds, *viz.*, that the Divinity might in other planets "resume a physical form, and expiate the guilt of unnumbered worlds." *

Source of difficulties in narrow views of Christianity.

These suggestions may be safely left without further comment. Minds so constituted as to feel such a difficulty will probably be well satisfied with the solutions here proposed. But the whole discussion cannot but suggest a passing remark on the style and tone of theologising evinced in the very statement, whether of the objection or the answer. It seems to belong to a somewhat obsolete school, and to refer too much to those narrow humanised ideas of the Divinity and His dealings with man de-

* Page 142.

rived so commonly from too literal an interpretation of the anthropomorphisms of the Hebrew Scriptures, and little consonant with the more enlightened views which a better dispensation encourages.

The same difficulties urged by ohn Wesley.

The same religious difficulties adverted to by both the writers in this discussion were urged with some force long ago by John Wesley *, who (like

* I am indebted to a friend for the following extract from a sermon of John Wesley, Text, "What is Man?" Psalm viii. 3, 4.

"Let us then fear no more! Let us doubt no more! He that spared not His own Son, but delivered Him up for us all, shall He not with Him freely give us all things?

"'Nay,' says the philosopher, 'if God so loved the world, did He not love a thousand other worlds, as well as He did this? It is now allowed that there are thousands, if not millions of worlds, besides this in which we live. And can any reasonable man believe, that the Creator of all these—many of which are probably as large, yea, far larger than ours—should show such astonishingly greater regard to one than to all the rest?' I answer, Suppose there were millions of worlds, yet God may see in the abyss of His infinite wisdom reasons which do not appear to us why He saw good to show this mercy to ours in preference to thousands or millions of other worlds. I speak this even upon the common supposition of the plurality of worlds—a very favourite notion with all those who deny the Christian Revelation; and for this reason, because it affords them a foundation for so plausible an objection to it. But the more I consider that supposition, the more I doubt of it: insomuch that, if it were allowed by all the philosophers in Europe, still I could not allow it without stronger proof than any I have met with yet."

He then cites the argument of Huyghens, to the effect of the probability of planetary inhabitants, but proceeds to mention that at a later period Huyghens entertained doubts on the subject from observing that the moon had no atmosphere. Hence, he says, the argument falls to the ground. He then goes on: "But, you will say, suppose this argument fails, we may infer the same conclusion—the plurality of worlds—from the unbounded wisdom, and power, and goodness of the Creator. It was full as easy to Him to make thousands of worlds, as one. Can any one, then, believe that He would exert all His power and wisdom in creating a single world? What proportion is there between this speck of creation and the Great God that filleth heaven and earth, while

"'We know, the power of His almighty hand
Could form another world from every sand?'

the author of "The Plurality") could find no other solution than by denying the existence of other inhabited worlds, as an unfounded idea, merely taken up (as he says) by infidel philosophers, as affording a plausible objection against Christianity.

But the whole objection turns on the same tacit but monstrous assumption, that because the privileges of redemption *are granted* to the inhabitants of this earth, they are therefore *not granted* to those of any other worlds: and that *it is* a part of Christianity to hold this exclusive view.

The objection depends on an unfounded assumption.

That such high privileges (as already remarked) are asserted with respect to man on this earth in Scripture is manifest; but it may be asked in vain what particle of proof can be alleged for *denying it*, with respect to *the possible inhabitants of other worlds?* Let those who urge this objection pro-

"To this boasted proof, this *argumentum palmarium* of the learned infidels, I answer, Do you expect to find any proportion between finite and infinite? Suppose God had created a thousand times more worlds than there are grains of sand in the universe, what proportion would all these together bear to the infinite Creator? Still, in comparison of Him, they would be, not a thousand times, but infinitely, less than a mite compared to the universe. Have done, then, with this childish prattle about the proportion of creatures to their Creator, and leave it to the all-wise God to create what and when He pleases. For who besides Himself hath known the mind of the Lord, or who hath been His counsellor?"

No proof of the exclusiveness of privileges.

duce a single argument from the reason of the case (if they think it one amenable to reason at all), or a single passage from the New Testament in which *such limitation or exclusion* is asserted, and the objection may have some weight. But this they do not attempt, nor do they seem to perceive that the attempt is necessary to establish their case.

Inconsistencies of Wesley's argument.

That Wesley should set down all philosophers who advocate a plurality of worlds as "learned infidels" is not surprising; but with his acknowledged acute powers of reasoning, it is remarkable that in earnestly asserting (what is not contested) that such high privileges *are* vouchsafed to man, and expatiating on their greatness and value, he seems to think it sufficiently proved that they cannot be granted to other races of beings.

And when he so strangely glances at, but passes over, the undeniable argument that *Omnipotence* could as easily *create* thousands of worlds as one, he seems equally blind to the obvious answer that the same Omnipotence could as easily *redeem* thousands of worlds as one, if they needed it.

Contradiction in reasoning on such a subject.

The truth is—all these and the like difficulties arise wholly from the inconsistency of *attempting to reason at all* on subjects which the writers themselves, at the same moment, pronounce *to be above all reason:* attempting to argue that the Deity could, or could not, act in this or that way, when they, in the same breath, assert the incomprehensibility of His counsels.

If it be an inscrutable mystery *wholly beyond human comprehension* that God should send His Son to redeem this world, it cannot be a *more* inscrutable mystery or *farther beyond* human comprehension that He should send His Son to redeem ten thousand other worlds. If, on the other hand, the *mystery* be amenable to any reasoning, or charge of inconsistency in the one instance, it must equally be so in the other.

But, apart from the *mere words* of the sacred writers, these doctrines have no existence; and these words express nothing with respect to the case of other races or other worlds; addressed to our own race, they profess to declare only what concerns that race.

Other instances of narrow views.

Though referring to a different part of the subject before discussed, yet I may just cite another instance of expressions evincing a similar narrow spirit, and seeming to imply ideas which might have been supposed exploded from the minds of philosophers: — "Can we believe that He who formed the *worlds* has made only *one*, and that in place of resting on the seventh day, He rested during the whole week of creation, and still rests, having transferred His almighty power to certain laws of matter and motion, by which the sun and all his planets were *manufactured* from nebulous matter?"*

The common occurrence of language of this sort (not solely in reference to the present question) leads us to reflect generally how much of the unhappy perplexities and objections which beset the minds of believers on some points, especially where science is supposed to come into collision with religion, must be traced to the influence of popular dogmas, founded on a narrow literalism, which, as in cases already glanced at, betrays its Judaical origin, rather than

* Brewster, p. 249.

any connexion with the enlightening influences of Christianity.

Peculiar views as to future state.

There is one other idea of an extremely peculiar kind taken up by Sir D. Brewster*, referring to the question *where* believers can place the *locality* of their future state. He refers to some passages of Scripture, which literally seem to imply, or at least countenance, the idea of a plurality of inhabited, or at least habitable worlds. He enters on a calculation to show that, in a future state, for the myriads of resuscitated human bodies, the earth would afford utterly insufficient room; and that the future abode of man must, therefore, be in some of the other bodies of the solar system, which, *being habitable,* will be suitable to this purpose. Such is the idea indicated even in the introduction of his volume, and such the final conclusion to which the whole discussion leads.

It is the danger threatened to *this* doctrine which constitutes the main cause of alarm at the triumph of scepticism in the denial of a habitable condition to the planets. Yet it might rather seem that their

* Brewster, pp. 18. 256.

being uninhabited would be more favourable to this doctrine, as affording more ample space for the reception of resuscitated humanity.

Difficulty obviated.

Though unable to perceive the importance or reasonableness of this question, I am yet anxious to give it the most respectful consideration; and therefore feel bound to add, if it be needed, for the confirmation of any wavering mind, that, in my opinion, the slightest attention to the writings of the Apostles affords a more satisfactory solution of the difficulty than any astronomical theories whatever. If there be one point clear in their declarations, it is that the resuscitated body will *not be material.* "It is sown a *natural* body — it is raised a *spiritual* body." (*σῶμα ψυχικον, ... πνευματικὸν.*)*

General remarks on the application of the arguments.

Such then, on the whole, is the formidable difficulty of a plurality of inhabited worlds! Such the dispute which threatens the alternative of a surrender either of faith or of science! Such the source of so much perplexity to thoughtful and religious minds, to the solution of which such elaborate

* 1 Cor. xv. 44.

speculations must be devoted! Such the danger threatened to Christianity, which it is the aim of the essayist, by such laboured reasonings and startling paradoxes, to avert! And to escape from which Sir D. Brewster, by so opposite a route, would guide his readers!

It is difficult, perhaps impossible, fairly to judge of the *convictions of others*, and I would wish to treat all serious convictions with unfeigned respect. But after what has been already observed, I must confess myself more disposed to concur in the abstract *justice*, than to perceive the *consistency*, of Sir D. Brewster's remark, "The difficulties we have been considering, in so far as they are of a religious character, have been *very unwisely* introduced into the question of a plurality of worlds!"*

Recommendation of clearer views of the subject in its first principles;

For my own part, I would rather disclaim the entire principle of such discussion, whether in the objections or the replies. The strange juxta-position of ideas of such very different kinds, religious and physical, which characterise these reasonings

* Brewster, p. 154.

altogether, seems in a peculiar degree likely to expose the whole subject to the attacks of the scoffer. The expression of theological dogmas contrasts singularly when mixed up with the speculations of science. The languages of the two sound strangely together; and I am powerfully reminded of the wisdom of Bacon's suggestion, "DA FIDEI QUÆ FIDEI SUNT."* And if these speculations on planetary worlds have really caused any perplexity to persons capable of profiting by rational and philosophical views of the subject, I would rather entreat their serious attention to the question, whether the cause of truth would not be better served by a preliminary endeavour to acquire clearer views of the grounds on which scientific speculation on the one hand, and the expression of theological doctrine on the other, are legitimately established, and of the very different basis on which they respectively stand, each firmly on its *own* ground, but as unstable on the other's as a ship on land or a pyramid on the sea. †

both in philosophy and theology.

* De Augm. lib. iii. c. 2.

† That it is not unseasonable or needless to press the consideration of the *proper ground* on which such inquiries ought be conducted, nor to insist

To this end, then, to aid in such an inquiry, a very few concluding remarks may be directed.

Some general suggestions on this subject.

The case as put by the essayist, "How to place our theology," with reference to the question of inhabited worlds, may perhaps be taken as in some sense the expression of a difficulty felt more widely as to the *relation of Christianity to physical science generally;* and is probably similar to that acknowledged by another eminent writer, who declares it to be "the great problem of the age to reconcile faith with knowledge, philosophy with religion." *

To discuss such a question in the way which its importance demands, would be to open a very wide inquiry, wholly beyond the necessary limits and scope of this Essay, and one which, perhaps, could hardly be entered upon with much prospect of satisfying the varieties of apprehension (or possibly misapprehension) which it might encounter. All I shall attempt, therefore, will be to offer a few gene-

on the independence of matters of faith from those of science, is further evinced by the announcement, while these sheets are going through the press, of another anonymous publication on the theological argument, "The Plurality of Worlds.—*The Positive Argument from Scripture*," &c. London, 1855.

* Archdeacon Hare, Life of Sterling, p. 121.

ral and somewhat fragmentary remarks, which must be left to the judgment of the reader to apply.

Plurality of worlds only a subordinate point.

In fact, the question now discussed is only a subordinate branch of a far wider subject. The mere inquiry as to the probability, or otherwise, of inhabitants in the planets, is in itself of a very secondary and unimportant character. What we have to consider is rather the broad principle involved in *any or all cosmical speculations;* they all tend to place our earth, and man upon it, in a very subordinate position in the vast universe; not merely in space and position, but in general relation; as one small and insignificant link in the vast chain or rather universally connected ramification of physical causes, of which there is no one constituent part more the *head* or the termination than any other. Hence a difficulty is felt by some because they imagine that moral and spiritual relations must follow the same law.

Subordinate position of man in the universe.

Yet nothing can be in itself more unfounded; and still more must this appear when considered in connexion with what has been here advanced. The tenor of the whole preceding argument has been to point out the *independence of the physical order* *of things, and the spiritual.** It has been maintained that the very idea of a spiritual nature in man, *in so far as* it is independent of the body, belongs to a higher order of conceptions, of a kind radically different from, and forming no part of, a physical system.

Spiritual truths distinct from physical.

It is to this class of conceptions that religious doctrines properly refer; and thus it would seem, *on general grounds at least*, unreasonable to expect that the two should have any connexion; or to be anxious, on the one hand, to frame theories combining physical science with religious belief, or, on the other, to imagine the region of physical truth an unsafe locality for a theological creed. But *in detail*, perhaps it may be said this will be found otherwise.

Connexion of science and religion in natural theology.

There is, certainly, one point of view in which physical science and theology are obviously and unquestionably in close contact and dependence—in the primary inference of a Supreme Intelligence, as derived from the order of nature so largely dwelt upon in the preceding Essay †; and which

* See above p. 74. 242.

† See Essay I. § v.

forms the substantial and necessary foundation of all *rational* conceptions of religious belief. But this *foundation,* however solidly laid, rises no higher than to the lowest basement; and if the conclusions at which *reason* arrives, are restricted according to the nature of the physical evidence, while they may afford some corrective of too blind and literal a dogmatism, they offer no disparagement to higher spiritual convictions. In this respect, then, and to this extent at least, we may, perhaps, see "how to place our theology" on the basis of these sublime deductions from physical science. These deductions (as before mentioned) are confessedly *very limited and imperfect.* They present nothing, as it were, but the meagre skeleton; to fill it out with substance and life is the function of those higher intimations derived from moral and spiritual sources, and which are, in their essential nature, alien from physical considerations.

Natural theology very limited.

Natural theology prepares for revelation.

A scientific natural theology does not rise to the aspirations of a spiritual or moral Theism; still less to the scriptural or ecclesiastical doctrines. It tells *very little* of the *attributes,* and NOTHING of the *mode* of the Divine existence; but for that very reason it presents *nothing to contravene* higher spiritual views *proposed from* OTHER SOURCES.

Physical difficulties in more particular religious views.

Yet, when from this primary truth, we advance to those more precise views of religious doctrine, we find a disposition continually evinced to place them in connexion or in collision (as the case may be) with physical considerations; to raise philosophical theories on a theological basis, or to find fatal difficulties in the failure of such attempts.

Sources of misconception on these subjects.

If we would trace such tendencies to their source, we may find a too common origin of misconception in the ignorance in which many, even of considerable scientific attainments, remain as to the real nature of Christianity; when their profession of it consists either in merely bowing to conventional requisitions, or is based on notions derived solely from the prevalent or established belief, instead of an enlightened and independent examination of it for themselves.

Ignorance of Christianity.

Want of sound philosophical principles.

Again, many who are extensively versed in the details of a *particular branch* of science, may often have reflected little on its wider relations and philosophical spirit. Hence, while they admit the impropriety of some of the speculations just adverted to, they are deficient in the distinct conceptions of the broad principles and grounds of all philosophical inquiry.

Applications of physical philosophy.

Physical philosophy has doubtless within itself the *germs* of higher knowledge, and presents us with those first elementary notions which are pre-eminently valuable, as subservient to the establishment of theological truth on a rational basis. In such a sense, and within due limits, *theology deduced from philosophy, may* be sound and valid. But in every case *philosophy deduced from theology must* be *essentially* erroneous and fallacious; it is no longer *philosophy.* It appeals to other authority, and disowns its proper inductive character.

Erroneous systems from misconception of first principle.

The desire whether for peopling or for dispeopling planetary or sidereal worlds *on theological grounds,* appears to arise from the same fundamental misconception or disregard of the proper provinces and limits of philosophy and of theology which has led, in so many other cases, to an unhappy and incongruous mixture of the two,—producing nothing, as Bacon has so justly observed, but "a fantastical and superstitious philosophy and a heretical religion." * Of

* De Augm. l. iii. c. 2.

this mode of procedure we have had abundant instances in all stages of scientific advance.

Theological philosophy.

Without recurring to more ignorant ages, and the speculations of the schoolmen, we trace the very same spirit in later times in the formation of such systems as that of Tycho, founded on the idea of reconciling astronomy and Scripture; in the vortices of Descartes, deduced by reasoning on *theological* grounds from the perfections of the Deity; in the cosmical theories of the Hutchinsonians, or what they termed "Moses' Principia," founded on the Hebrew Scriptures, in opposition to Newton's; and in our own times in the various schemes of the Bible geologists, each in succession presenting but some new shade or modification of the *same radical misconception* to take the place of its exploded predecessor.

Bible geology.

It is worth while to dwell on this last instance as very instructive in its consequences, especially to those who have not antecedently taken more general views. Even at the present day there are not wanting occasional attempts to keep up the hopeless chimera of erecting theories of geology on the Mosaic narrative. It is needless to observe that, as all notion of an accommodation of the *facts* to the text has long since been given up by all *sane* inquirers, these attempts are now merely directed *to explaining away* the sense of the text; in which they no doubt succeed by *such* principles of verbal interpretation as, if fairly applied to other parts, would readily enable us to put on any given passage any required construction.

Contradiction between geology and the Hebrew Scriptures.

All inquirers, possessing at once a sound knowledge of geology, and capable of perceiving the undeniable sense of a plain circumstantial narrative, now acknowledge that the whole tenor of geology is in entire contradiction to the cosmogony delivered from Sinai; a contradiction which no philological refinements can remove or diminish; a case which no *detailed* interpretations can meet, and which can only be dealt with as a whole.*

I have elsewhere† fully discussed this subject, and have there explained the only view which I think the case admits; in one word, that the narrative, as a whole, as it cannot be received as *historical*,

* For some excellent remarks bearing on this point, the reader is referred to Mr. Kenrick's "Essay on Primeval History." London, 1846. Preface, p. xiv.

† See Connexion of Natural and Divine Truth, p. 245.

may be regarded as a *poetical*, representation; adapted, as it was addressed, to the Israelites, as the basis of the institution of the Sabbath. But be this as it may, real Christianity, I contend, can be in no way affected by this, or any, contradiction to the Old Testament law *, with which it has been erroneously mixed up; on the contrary, the palpable discrepancy is valuable, as reminding us the more forcibly of its independence.

Encroachment of science on erroneous belief.

It is undeniable that the advance of physical knowledge has from time to time made inroads on the territories which prescriptive error had once consecrated to religion. So the Copernican heresy not only deposed the earth from its proud immobility as the centre of the universe and the throne of spiritual infallibility, but set at nought the *letter* of numerous scriptural texts; it entailed the impious doctrine of *antipodes*, and destroyed the ideas of *an upwards* and *a downwards*, a *local* heaven *above*, and a *local* hell *beneath* the earth. It broke through the solid firmament, and placed in jeopardy the existence of a physical purgatory. Yet real Chris-

* See my Essay on the Law and the Gospel, "Journal of Sacred Literature," April, 1848.

tianity has been in no way injured; but the reverse; its rational and spiritual character has been the more powerfully asserted and vindicated.

In all ages bigotry has erected its strongholds on the basis of ignorance, and especially on erroneous physical ideas, and has then resisted all advance of intelligence on the plea that it is destructive to the security of religion, as it doubtless is to their baneful systems; while they thus by clear implication dissever the claims of religion from those of truth, and degrade the profession of it to the level of the most baseless superstition.

Favourable to a purer religion.

Even among professed Protestants, and in a philosophic age, men have hardly become convinced that the advance of physical enlightenment, so far from being hostile to religious truth, is eminently serviceable to it, were it only in dissevering it from false allies and equivocal auxiliaries, and thus exhibiting its true spiritual power, when cleared from the heterogeneous incumbrances and corruptions which a false philosophy or a narrow literalism had fastened upon it, and which, instead of aids and defences, are in reality its hindrances and disfigurements.

But when we turn to the pages of the Bible, it is doubtless the fact that continual reference is made to *physical considerations* of various kinds, which may in many instances give rise to difficulties. These will be so very differently estimated in magnitude and significance by different minds, that it would be impossible to discuss the question in a way satisfactory to all. It may perhaps suffice to suggest a general application of what was above observed in reference to the instance of the geological discrepancy.

Physical difficulties in the Bible.

Whenever the sacred writers introduce *physical* statements, they may fairly be understood as speaking *conformably to the existing state of knowledge*, or adapting themselves to the ideas, belief, and capacities of those they addressed. In any such cases it would be irrational for us at the present day to insist too literally on such representations, and especially to reason on them in cases where we are precluded from examining into all the circumstances, or canvassing the evidence. But if after all, in any instance the letter of the narrative or form of expression may be found *irreconcileably at variance with physical truth*,—without trenching on the integrity of the text, or

General rule for physical difficulties in Scripture.

of the testimony,—we may allow the alternative of understanding them in a *figurative or poetical sense.*

Objections from want of distinction between the different parts of the Bible.

Difficulties of this kind in many instances, especially as regards the Old Testament, derive their whole *importance* from the common want of due discrimination as to the distinct object and character of the different portions of the Bible.

If, however, we look with a more discerning eye to the nature of its contents, in the first instance we find a record of older and imperfect dispensations, adapted, as they were addressed, to the ideas and capacities of a peculiar people and a grossly ignorant age—a law of "carnal ordinances" and sabbaths, specially founded on that peculiar cosmogony which we now know to be untenable; physical influences, temporal and national retributions.

Christianity independent and spiritual.

But the more perfect and universal religion of Christianity, if in its first outward manifestations accommodated to the convictions of the people among whom it originated, yet in its essential characteristics and more full development to the rest of the world, as set forth in the Pauline Epistles, while it expressly disclaims the peculiarities of the older dispensation, exhibits characteristics of a higher, more comprehensive, and spiritual character; professedly appealing for their acceptance to the principle of *faith*, not of *sense*.

Doctrines of the Divine nature.

If we look to its more special doctrines, as *regards the Divine nature*, we may observe that a physico-theology supplies *no such idea of the Deity* as can offer any antecedent contradiction to the representations of His nature and attributes (necessarily more or less anthropomorphic), or to the spiritual mysteries of redemption and a future state, *as announced in the language of the New Testament.*

Of the nature of man.

And as to *the condition of man*, the language of the Christian doctrine represents it not in reference to any material, metaphysical, or moral hypotheses, but to peculiar spiritual principles. It makes the spiritual man a distinct being from the natural,— "*a new creation;* and is engaged not in tracing *physiologically* the *origin* of the human race or of its *infirmities*, but *in providing the* REMEDY. It does not dwell on external events in any physical detail, but always with a doctrinal application, or in a spiritualised meaning. Its *essential* design belongs to *spiritual* things; its relation to external and *physical* things can be but *incidental;* and of the proper objects of its revelation we may truly say with Bacon, "*Dignius credere quam scire.*" *

Christianity independent of physical theories.

Thus we need be in no trouble "how to place our theology" among physical systems; nor need it fear any speculations as to the inhabitants of other worlds or possible revelations granted to them. It has no concern with chronology, astronomy, or cosmogony (least of all, that of the Judaical law); with the nebular origin of the planets, or with the development of successive races of organised beings in them or on the earth; with the myriads of ages which mark the antiquity of the world, or the date of man's origin upon it; or with the question of his derivation from one stock or many.† It leaves these questions to be guessed at as they may, or investigated on philosophical principles so far as they can be. Its peculiar aim is entirely different and

* De Augm. lib. ix. c. 1.

† While writing this Essay, I have received a copy of a pamphlet, circulated privately and anonymously, in which an able and learned writer, evidently a strict upholder of inspiration, endeavours to show by elaborate, critical, and philological investigations, that the Bible distinctly sanctions and asserts the idea of the primeval existence of other races of men besides the family of Adam. It also includes an able defence of this belief against its supposed unfavourable influence on the doctrine of original sin. ("The Genesis of the Earth and of Man," &c. printed for private distribution, July, 1854.)

independent: its objects *belong to another order of things;* and its representations of them are avowedly *not the realities,* but only their *images;* they can be seen by us only, "*δἰ ἐσόπτρου ἐν αἰνίγματι* *,—by means of a mirror and in an enigma," in our present state; while it holds out a future, when "we shall see face to face, and know even as we are known."

* 1 Cor. xii. 12.

ESSAY III.

ON THE

PHILOSOPHY OF CREATION.

THE PHILOSOPHY OF CREATION.

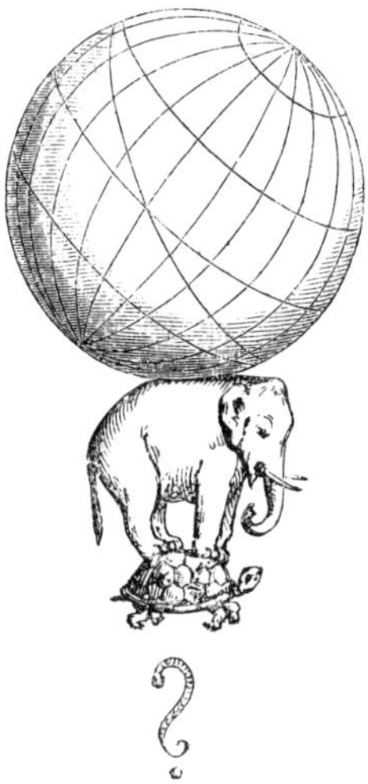

INTRODUCTION.

THE question of "creation,"—whether in the higher sense, of the first origination of the material universe, and of all physical causes,—or in the secondary and more accessible meaning, of the earliest history of the cosmical arrangements of stellar or planetary systems,—or more particularly of our own globe, of its physical revolutions, or of the successive introduction of new forms of organised

Preliminary remarks. Interest of the question.

life on its surface, — has at the present day excited an unprecedented degree of general interest.

Various grounds of discussion.

And the discussion which has been called forth has exhibited the greatest variety of tone and character. Though by some the subject has been viewed in a sober philosophical aspect, yet by others it has been made the subject of hypothetical speculation; sometimes running off into what must be deemed very fanciful imaginations, or occasionally involving metaphysical ideas, carried out into various stages of abstruse and even mystical speculation. By others, again, it has been taken up on religious grounds, or mixed up in various degrees with the influence of a theological creed, to which speculations of a more scientific kind have been deemed hostile.

Religious bearing of the subject.

That the subject has a direct connexion with such higher considerations is manifest; but this by no means implies such a confusion of ideas as would vitiate the claims of independent philosophical inquiry, or sanction the attempt to found conclusions relative to the physical order of things and the structure of the material world on a basis totally alien from that of *inductive science.* The truth is, the same observations will in a great degree apply to this question which were offered in a former Essay*; and the strange speculations which have been sometimes broached respecting it have probably originated in a great degree in the want of clear appreciation of the distinct grounds on which our convictions of scientific and of religious truth respectively should be based. And until these distinctions are properly drawn in the mind of the inquirer, it will be to little purpose to discuss the details of controversy.

Properly a subject for physical conjecture.

It will be allowed, then, in conformity to principles before laid down, that on purely physical and inductive grounds it is fully open to us to inquire how far science can legitimately conduct us towards some indications of the *mode* in which, and the *secondary processes* by means of which, the first establishment of the existing natural world may have been worked out, and limited strictly to the tenor of recognised natural analogies, to speculate on the probable order of evolution of the earliest rudiments of life.

In proportion as such speculations have a tendency to impress a conviction of truths of a higher order,

* See Essay II. § II.

it is the more necessary to the force and validity of such inferences, that the *evidences* on which they rest should preserve a free, independent, and unprejudiced character, and should not involve an assumption of the points to be proved.

Meaning of the term "creation."

It must also be borne in mind, that we have by anticipation, in several parts of the foregoing Essays, discussed more or less in detail several topics which have a material bearing on the present subject, as referring to the past history of the world, and those stages which it has passed through in the process of formation; all which are essential to any *physical* view of the nature of its *creation*; that is, its history, *so far as we can trace it*, towards its first origin.

The very use of the term "creation" may indeed be supposed to point to associations of a higher kind, which are altogether beyond the simple scientific question. But, in a philosophical sense, it should be carefully borne in mind, that if that term be employed, we can regard it as no more than an expression of our ignorance as to the *mode* of the first origination of the material world; while as to the secondary points connected with that question, we may always look to the further enlightenment which discovery will continue to throw upon them.

Physical evolution.

Even in earlier times, and under very erroneous systems of philosophy, we trace some recognition of these sounder principles. Amid the many extravagancies and radically erroneous principles of the philosophy of Descartes, we may yet recognise its principal merit in that, in an age when *metaphysical* abstractions and causes alien from natural induction were universally resorted to for the solution of *physical* phenomena, his theory of vortices (imaginary as it was) yet at least referred to *conceptions and modes of action of a properly physical kind:* and it is remarkable, that in conformity to the same broad principle, he likewise distinctly upheld the notion of the origin of the existing *organic* world, evolved according to a series of regularly adjusted laws out of its primitive elements; though like some other *very prudent* philosophers, from a salutary apprehension of popular odium, he was unwilling openly to avow such an obnoxious tenet, and in his popular writings professes to think it more likely that the whole was created at once as it now stands.

In carrying on this inquiry with special reference

to some of the theories started at the present day, I propose, first, to consider briefly the general amount of information which can be regarded as bearing on such a question, furnished by those branches of science most directly connected with it, and then to offer some remarks on the general character of the reasoning raised upon that evidence, the kind of conclusions we may safely deduce, the kind of hypothetical speculation in which, if so disposed, we may legitimately indulge, and the extent to which any real conceptions can be carried of a subject which, in its entire compass and highest meaning, must necessarily be beyond the reach of positive investigation, or even of human comprehension.

§ I. — THE EVIDENCE DERIVED FROM GEOLOGY.

Sources of information. Fossil remains.

In any question as to the origin of the world pursued on scientific grounds, the sources to which we can look for any positive or substantial information, must, in an especial degree, be those opened to us by geology and palæontology; and of some of the most material facts and admitted theoretical opinions in these departments bearing on the question, it will be necessary to take a cursory review.

Theories of progression and non-progression

When we trace backward, by the light of fossil remains, the succession of varied forms of organised existence which have tenanted the surface of our globe during the incalculably vast periods of past time, the fact of their presenting apparently very different characters in different epochs naturally led geologists and naturalists to speculate on the question, whether those variations could be reduced to anything like *a determinate order or law of succession;* and probably the most prevalent opinion has been that, at least in a general sense, there has been a succession in the order of *progress or advance* from lower and more simple, towards higher and more complex forms of structure and function.

More recently, however, this has been much disputed. Not only have particular instances supposed to invalidate this law, been brought forward as demanding certain modifications in the statement of it, but the entire principle has been contested and positively denied. And those who have pursued the inquiry (*restricted merely to the question of fact*) have been ranked under the two schools of Progression and Non-progression.

It may, indeed, be fairly questioned whether such distinct designations, applied without qualification, can be fairly supposed to characterise any parties in the scientific world, or whether we may not rather regard the differences as of a more limited character. At any rate, a very brief summary of the principal arguments on either side will materially conduce to our object.

Arguments for non-progression.

On the one side, Sir C. Lyell has supported his views by insisting on the merely *negative* character of the evidence we possess as to the non-existence of many species in early ages; * on the *absence* of any indications in some of the earlier formations, from which we can form any idea of what the entire flora and fauna of those periods really were. More precisely, those formations being wholly marine, we have only evidence of the marine organisms, and can expect none of the contemporaneous land productions; and marine plants and animals are confessedly always of lower organisation. Throughout the long period of the Silurian formations, he contends that we have little evidence of any advance or progress in the scale of organisation. The fishes of the coal formation are of higher organisation than any existing species — in fact, combining reptilian characters with icthyic; and even up to the later formations he conceives that there is but little indication of any real advance in character.

Absence of evidence.

In general, that the remains actually preserved in any formation furnish us with no certain standard or adequate representation of the entire existing state of the organic world at the time, is evident from the very partial, local, and accidental manner

* Address to Geological Society, 1851.

in which those remains have become imbedded; their preservation has been the *exception* rather than the *rule.* We ought not to expect to find evidence of the possible multitudes of co-existing species which might be so circumstanced that their remains were never likely to be thus embalmed for our instruction.

No proof of inferiority of earlier fauna.

On such grounds, then, it is concluded that we have no real proof of the *general inferiority* of the organic world in the earlier periods, or of any superiority, at least through a long course of succeeding ages; nothing to lead us to trace backwards any determinate series which points to a primary rudimentary condition, or to the origination of a *more perfect* state of things out of a *less perfect.*

Arguments for progression.

On the other side, the argument which has been so ably sustained by Professor Owen*, turns chiefly on the *positive* evidence supplied by the comparison of those organic remains which *are* preserved to us in any one formation, with those in another, when, in fact, the same probabilities of preservation must be supposed to have subsisted, and yet the remains

* See Quarterly Review Sept. 1851.

of each exhibit such marked relative characteristic differences, that thus, he conceives, the inference of considerable progressive advances may be fully justified. But his main conclusions are supported, not so much on general arguments, as on minute anatomical comparisons; and such comparisons, in at least very numerous instances, present undoubted marks of physiological changes, clearly progressive according to the order of higher organic development.

Comparison of preserved forms in different periods.

But in some instances, it is further urged, we have a more positive ground of inference of the real absence of particular species, when we find other forms closely allied which are, by analogy with the rest of the system, fairly considered as the analogues or *representatives* of the missing species. We infer, as it were, the absence of the *principals* from the presence of their *substitutes*. The argument, of course, depends on the force and correctness of this analogy, and the general evidence of such a system.

Analogies of species.

With regard to the absence of terrestrial remains, it is considered by some as evidence that, in the earliest periods, the sea covered the whole surface of the earth, or nearly so. But there must have been

Marine formations.

some elevations to be washed down, and form the deposited beds.

Changes chiefly in higher forms.

Again, if progression be supposed, it is an admitted part of the law that in the *lower* forms, whether of animal or vegetable life, the change has been throughout all the series of eras always much less than in the higher, and that some of the lowest forms are persistent, or nearly so, through all formations. Thus it would be in proportion as we ascend to the higher classes, that any marked signs of change or improvement might be expected.

Influence of terrestrial heat.

There is one material consideration the force of which, *on any view of the question of progression*, it is impossible to overlook—the fact of the *central heat* of the earth, with its undeniable consequences.* A hot body in free space must cool; and if now cool at the surface and hot within, the earth must have cooled from a hotter state, and must once have been intensely hot; and, by the same rule, once in fusion, or in vapour. Here, then, there *must* have been a series of *progressive changes*; and if the cool-

* See especially Mr. Hopkins' Address to the Geological Society, Anniversary, 1852.

ing process *had not reached* its present apparent state of equilibrium *before* organised life began, it would be a natural consequence that some marked changes in animated nature could not but have accompanied those changes in temperature, and have followed a like determinate order. Whether the present equilibrium had been attained before the commencement of life may still be a question; yet, considering the enormous length of time through which organisation has certainly existed, it is difficult not to suppose that some part of the series at least must have reached back into the period of perceptible cooling, so that influences of terrestrial temperature may have been not without their effect on the changes of species.

Supposed local causes of heat.

On the other hand, these considerations may admit of qualification, if we should agree with those who ascribe the terrestrial heat wholly or partially to other causes than simply the remains of a primitive high temperature. If there be admitted, for example, any internal cause of combustion capable of being more or less excited from time to time, any effects of a progressive kind must be greatly influenced; and the advocates of the chemical theory of volcanic action allege much probable evidence in support of such an idea.*

Again, possible changes of temperature from external cosmical causes might interrupt those due to progressive cooling. Nor need we dwell upon those more immediate local causes of change of climate in the variations of physical feature in the continents and oceans, universally recognised by geologists, which might interfere greatly with any general progressive change of temperature in particular localities.

Influence of temperature on organic life.

Another question might arise:—Are the organised productions, at epochs when a tropical climate prevailed in districts now temperate or cold, upon the whole of lower organisation? or are we to infer that a hotter temperature is less favourable to the evolution of higher forms, or a colder more so?

In one word, we are not certain as to what were the *successive order* of changes in the temperature of the earth; nor if we were, could we thence argue what kind of successive changes in organised life should be expected to accompany them.

Upon the whole, when we carefully examine all that

* See Daubeny on Volcanoes, p. 430. (Ed. 1826.)

has been alleged on either side, we cannot deny the evidence, in some sense, of progressive changes on the one hand, though even throughout some long periods we may allow the amount of change is small, and the apparent amount greater than the real. All advance is at the utmost extremely slow; and with the progress of discovery, there is continually increasing reason for believing it slower than has been imagined, and that a high type of organisation prevailed in epochs much more remote than has been supposed; while almost every fresh discovery tends to push backwards the boundary which seemed to mark an inferior order of things into remote depths of primeval time. In fact, the discussion of the question of progression or non-progression is perhaps less valuable on its own account than as it leads to a more searching review and analysis of the ground on which all geological reasoning proceeds.

General bearing of the question of progression.

We are thus led to consider generally the proper force of *negative evidence* as such: in other words, that any inference made from the *mere absence* of discovered instances must be *essentially dependent on concomitant circumstances.* Negative evidence by

Force of negative evidence.

itself is simply neutral; but it acquires a different character and force, according as other arguments concur with it or otherwise. Negative evidence is *strong* in proportion as we may be able to show from circumstances a high probability that instances would be found if they existed, or a high probability, from other analogies, that they did not exist. *Non-appearance* would here be nearly tantamount to *non-existence.* But negative evidence is *weak* in proportion as we may be able to show from circumstances, a probability that instances would *not be observed* even though really existing; and *still weaker,* if analogy should render it likely that they did exist. *Non-appearance* would here be no presumption even of *non-existence.*

Origin of life on our globe.

As to the great question of the *first origin of life* on our globe, geology can give us very little, if any, information. An *azoic rock* is no necessary proof of an *azoic period.* Animals may have lived and flourished abundantly where, from peculiar causes, none of their remains might have been imbedded. Analogical considerations may avail to guide conjecture to a certain extent. But the earliest forms AS YET *known* are *not* of the lowest organisation. And if we descend to the so called primary rocks, and to those called metamorphic, it is clear, whatever remains they might have included, must necessarily have been fused and burnt up. In the inductive prosecution of the question of the progress of life, the first object would be to endeavour to determine the *law* to which the order of *succession* of species in different epochs may be found to conform. The first and most simple idea of a direct advance in successively higher organisation in one line, from the lowest zoophyte up to man, as we advance through geological ages up to the present time, is now acknowledged to be untenable; but what is the real order which we are to substitute for it, is not so easily apparent.

Law of succession of forms.

The assertion is often dwelt upon with a very mistaken emphasis and importance, that a particular species is *highly organised*, when the real point of distinction should be, not its *absolute* but *relative* degree of development; when the question is, not whether its structure is actually of a complicated kind or exquisitely adapted to the conditions of its existence (of which in no case is there any doubt), but whether it is of a *higher* or *lower* grade *relatively* to other creatures of a corresponding class in other formations.

And, again, it is a point of material importance, but one in which few writers are agreed, what we mean by *advance* or *progress*, or what really constitutes a *higher* or more perfect organisation.

Scale of advance.

It would rather seem that each species is higher in some respects, and lower in others; or that there are *many scales of perfection in different respects*, running, as it were, *parallel* with each other; and that in defining the degree of elevation of any species, we must take into account the position it occupies in the several different scales jointly.

Among existing animals, it is now generally allowed that the arrangement of species in the scale of organisation is not that of simple ascent in any line, or even in several branching in any one direction; it is more properly compared by Professor Owen* to a "net-work;"—every species being connected with others by a variety of ramifications, and not simply by ascent or descent in a scale.†

* Lecture, British Association, Liverpool, 1854.

† The same difficulty in the definition of *higher* or *lower* organisation has been felt also by other naturalists. That species may be higher in certain respects and lower in others, is also dwelt upon and illustrated

And as to the law of changes of species in *past epochs*, in the very imperfect knowledge we at present possess of it, it is at least clear that, so far from a regular *advance* from lower to higher forms, in many instances there appears rather a deterioration and degradation of character in the progress of time towards the existing state of things. But what seems most material towards the probable ultimate enunciation of a more true and general expression of the case, is the law of *combination and separation* of character; that is, a combination of the characteristics of several species, or even genera, or orders in the same individual in one period, to be developed separately in different species in a succeeding era; and this in such distribution as to present appearances of advance in some respects, along with degradation in others; as if in the functions of vitality, the principle of "division of labour" had been gradually introduced. Of this combination of charac-

Not a simple advance from lower to higher.

Combination of characters.

Afterwards separated.

Examples.

---

by Professor Pictet, "Traité Elementaire de Paléontologie," &c. (See Geol. Quarterly Journal, No. V. p. 50.)

And Oken, in proposing his new scheme of classification founded on the modifications of the organs of sense (Physio-Philosophy, § 3065.) confesses the difficulty of arranging animals on any satisfactory principle in the relation of *higher* and *lower* organisation. (§ 3561.)

In his own system, he expressly notices that, while each class stands above another, yet *in each* the lower animals are inferior to the higher animals in the next below. (§ 3582.)

ters in an individual species, examples are familiar to every student of geology. The sauroid fishes generally furnish obvious and abundant instances, and we may take as a single case which has been much dwelt upon, the *jaw* of the Asterolepis; in which a row of small fish-like teeth are combined with another set of the large reptilian form, or as Mr. Miller so forcibly expresses it, we find "the crocodile lying intrenched in the fish." At the same time its general organisation was of the inferior type of the cartilaginous fishes, having external plates resembling those of the recent Lepidosteus and Polyptemus, and a spiral coprolite indicating a visceral structure like that of the ancient Icthyosauri and the existing rays and sharks.*

Indeed, it might perhaps even be conjectured to be more like the general law, that this kind of *combination of the characteristics of higher, with those of lower, classes,* might be the distinguishing feature of all the earlier stages of animal life; and that the higher we ascend in time, the more we might expect to find types combining characteristics of

* Footprints, &c. 80. 104.

several (perhaps of all) classes, thence afterwards to diverge in distinct directions.

Principle of the continuity of geological phenomena.

One of the most material points in the whole inquiry relates to the question of *continuity of character* observable in paleontological indications throughout successive formations. To a great extent such continuity is on all hands admitted as marking at least large portions of the series of changes presented to us; but an important question arises respecting *interruptions apparently occurring* in that order and gradual succession of forms, on which considerable difference of opinion has prevailed. On this point, then, we must make a few observations.

Successive subdivision of all formations.

In the first place, the general tendency of all geological discovery has been, and continues to be, to break up large divisions into smaller; to obliterate sharp lines of demarcation by subordinate gradations; to subdivide formations; to trace intermediate deposits, lost perhaps in one locality, but detected in another; and thus its course continually tends to fill up breaks, to render the series more and more connected, and to confirm the belief in a real continuity of geological phenomena; though we may, *as* yet, be very far from realising it in all instances, or throughout all the series of changes.

Instances. All geology is full of instances of such progress. It was not many years ago that the whole mass of rocks below the old red sandstone, and above what was called the primary, was confounded together under the common name of "grauwacke." All the strata, again, above the chalk were alluvium or diluvium, London clay and fresh-water beds.

But we have now, in the one case, through the combined labours of Sir R. Murchison and Professor Sedgwick, the vast mass broken up into the well-marked serieses of the Silurian (including the Cambrian) rocks, with their several subordinate divisions and accompanying beds; while some of these are again in process of undergoing still further analysis; as in the researches of M. Barrande: again, in the other case, the labours of Sir C. Lyell and his later coadjutors, in the first instance reduced the chaos into order, by the grand divisions of Eocene, Meiocene, and Pleiocene; in their turn subsequently broken up into an increasing number of minor distinctions of older and newer Pleistocene, and Post-pleistocene; no doubt, eventually to be still further marked out by yet more minute shadings of difference in epoch; and thus indicating in every instance a more gradual succession and closer approximations in the affinities of species.

Unity of plan of organisation in all epochs.

Throughout all formations, the grand truth to which every accession of geological discovery bears witness in a more remarkable manner, is the principle of unity of plan continually exemplified in all the varieties of organic structures disclosed. Even the most seemingly monstrous and incongruous forms of animated existence in past times are all, without exception, constituted according to regular modifications of a common plan, and with parts, organs, and functions related by the closest analogies to each other; so that no sooner is a new specimen detected than it immediately finds its proper position in the scheme of nature; no sooner is a new form discovered than it is instantly assimilated with some known type, and found to hold an assignable place in the system. Whether a given organic fossil (as in some instances in more recent beds) exhibit characters differing from some known form only as a variety or sub-species, or whether (as in earlier cases) it present features unknown to any existing genus or order, or (as in other instances) offer conditions in any degree intermediate, still in all cases alike the remarkable point is always, that a place and a name can be immediately assigned to every new form as it presents itself; and this too invariably in such a manner that it either tends to supply a link in affinity between orders of beings already related, or indicates some new and unexpected point of analogy. There is never any deviation from system and regular plan; we never light upon a fossil centaur or palæozoic mermaid; there never occurs any junction of heterogeneous members, any real departure from type and system. The invariableness of the results through such enormous series of ages cannot but impress the mind, when duly considered, with the highest idea of the preservation of continuity.

Connexion in affinity of forms.

Throughout all the most recent formations, indeed, we find a continuous series of allied species, and a succession of organised structures, in a chain absolutely unbroken, and marked only by the minutest specific differences in its successive links, down to forms now existing; and as this is carried backwards through countless ages, by degrees we find fewer features of the present, and more of the past, and even come to whole genera, and orders of extinct races coexisting with some which have survived them. But in some instances, especially in the more ancient formations, the series of forms present a more fragmentary appearance. At intervals in the course of this series of close and continual connexion, there are real or apparent interruptions of greater or less magnitude, in which the immediate affinity seems broken off between the species characterising one formation, and those nearest allied to them in the next formation.

Some apparent breaks in the series.

The case of apparent breaks or discontinuities between one great group of formations and another is often alleged as one main difficulty in the way of any theory of continuity; and this is evinced in several marked instances in an apparent interruption, not merely in species, but even in genera; that is, though through considerable ranges of closely consecutive, deposits, the transition of species takes place by insensible gradations, yet, at length, we come to a broadly marked separation of that group from another, where not only the species, but even the genera disappear and are replaced by others.

Examples.

To take a single example; one of the most remarkable of such apparent interruptions is, that marking the boundary between the Permian beds—the highest of the older group of fossiliferous rocks—and the Trias, or lowest of the next series; and to which respectively the terms Palæozoic and Mesozoic have been applied, as evincing an apparent entire change in the organic life which prevailed in those respective periods.

Sometimes a stratum containing, perhaps, abundance of fossil remains of a particular class or epoch, is succeeded by a great thickness of deposit totally devoid of organic remains; and then in the next bed below this, organic remains shall again occur abundantly, but of totally different *species*, or perhaps even *genera*, from the last, which, it is contended, indicates the occurrence of a long interval of time after the destruction of the former, and followed by the introduction of the latter kinds of beings, without any intervening links in the chain of existence appearing.

Thus Mr. H. Miller dwells with peculiar emphasis on an instance of this kind occurring in Orkney, where there is a bed of the lower old red sandstone containing an abundance of fossil fish; greater (according to this author) than in all other formations together; in which the celebrated Asterolepis was discovered.

Superimposed on this are other beds of sandstone, reaching to 1500 or 1600 feet of thickness, in which not a single organic fossil has been observed; above which are beds containing totally different species. Again, * he argues on the fact that in the Silurian system, fossil fish (at the time he writes) have been discovered only in certain beds in the upper division; while the lower, more than 3000 feet in thickness, are destitute of them, and below these we arrive at remains of a different character.

Objections against continuity.

These phenomena, and others of the same kind, have been the subject of considerable dispute; and the opponents of the doctrine of continuity, chiefly on grounds which it is difficult to recognise as connected with those of true science, and often in a tone still less reconcilable with its proper spirit, have been fond of triumphing in these facts, as if they inflicted a fatal blow on the views of their opponents. It is not my intention here to descend into any such polemical disputes. I merely proceed to the philosophical consideration of how far any such phenomena (granting the representations made of them as accurate)

* Footprints, pp. 106. 114.

General view of the evidence.

really affect the question of continuity. It will here be important to recur to the consideration of the nature of negative evidence; and to observe that the *absence* of all organic remains in a particular formation or deposit is *no proof whatever* that animal life did not abundantly exist during the whole period of that deposit, but merely shows that local and other causes did not favour the imbedding and preservation of their exuviæ.

Aqueous deposits.

The popular apprehensions as to the nature of geological events are often very inadequate and confused; and it is a point apt to be overlooked, that the terrestrial remains in all formations are merely *indications of what was the state of the* LAND *left us by the* WATERS, whether of the ocean, of rivers, or of lakes. Such remains were only occasionally imbedded—"rari nantes in gurgite vasto"—and thus afford no adequate representation of terrestrial life. Even marine remains are far from affording a complete memorial of the inhabitants of the ocean. At all events, it is a hazardous process to frame *theories* on the *absence* of such remains.

Submergence.

Exceptions may, indeed, be conceived in cases where, instead of being formed by sedimentary deposition, a tract of land, with its plants and animals, may have been *submerged by subsidence;* here a fairer representation of the whole fauna and flora might be expected: but, perhaps, very few, if any, instances affording good evidence of such a process have been clearly established.

All deposits local and occasional only.

Again,—all the formations which geology has traced were simply *local* and *occasional* deposits, extending sometimes over a greater, sometimes a smaller area; and going on at one time, and ceasing at another. Equally local, too, was the diffusion of organic forms.

Professor E. Forbes * has justly observed in his able comment on the labours of M. Barrande in the Silurian formations, "Thus early in the world's history do we find the partitioning of the earth's surface into natural history provinces; more and more evident does it become every day that the old notion of a universal primeval fauna is untenable." And to the same effect I must refer the reader to some profound observations of Mr. Darwin †, into

* Address, Geological Society, 1854, p. 33.
† Geology of South America, p. 105.

which my limits alone prevent entering here at the length they deserve.

At any one epoch deposits might be going on with more or less regularity during a certain period, enclosing remains; afterwards, during another period equally long, or much longer, a *total cessation of deposition* might take place, owing to changes in local condition; or, again, at one place *deposits* might be going on along a shore, while over a vast region, away from the waters, species of terrestrial animals might be flourishing in profuse variety, not one fragment of whose *remains* might ever be *washed down* or *imbedded* either in lacustrine or marine beds; in a way, in fact, exactly analogous to what is going on at the present day.

Alleged break between the earlier formations.

Again, it is alleged that the change from one great group of formations to another, at least in several marked instances (as *e. g.*, in passing from the palæozoic to the mesozoic period), was marked by the occurrence (according to some) of a "great convulsion," or at least of very extensive changes in the physical order of things, of which the condition of the strata sometimes bears striking evidence. Now, granting such changes as great as the catastrophist may imagine, it is surely a most unreasonable inference that these changes were such as to *destroy* all the species existing during the previous formation, and thus to leave the surface wholly untenanted until a new order of things supervened, and a totally new introduction of life took place.

Lapse of time between successive deposits.

But, granting that between the periods of formation of the upper and lower groups referred to, great changes in physical arrangements took place, it would be far more accordant with all reasonable analogy, in proportion to the magnitude of those changes, to allow a corresponding lapse of *time;* which being unmarked by any depositions, giving evidence of its duration by the successive changes they might exhibit, would necessarily remain to us a blank; a period which the advocate of natural causes may, with just as much probability and confidence, assert to have been enormous and incalculable, as the catastrophist can maintain it to have been brief and spasmodic.

If an interval of unknown and incalculable length intervened between two recognisable formations, and during all this vast time circumstances did not allow the imbedding of any characteristic exuviæ, it would be utterly vain and futile to assert that there was necessarily any breach whatever of *the law of continuity;* or to affirm that, during the whole of this enormous period, of which we are, from the conditions, precluded from knowing anything, all the species of the earlier epoch were not continuously existing and as slowly changing (by whatever means or law) to others more and more different, along with corresponding slow changes in physical conditions, until at the period when things were such that remains were again deposited, the whole character of the fauna had changed in the manner observed.

Analogy with those cases when deposits are are continuous.

In a word, in all those geological periods during which we *can* trace a continuous and gradual succession of formations without marked or violent interruptions, there we invariably find a like slow and gradual change of animated life, proceeding by small modifications of *species*, until, at length, comparing the extremes of the series, whole genera may be changed. If, then, in certain other cases, we find apparent *interruptions in the order of species,* apparent breaks in this orderly succession, or between such deposits of so different a character, periods intervening, during which we see that great changes or disturbances were in progress, as we must infer that those changes went on by the regular operation of physical laws, exactly as in the cases in which we *have* uninterrupted evidence—so, by parity of reason, we must infer that the like gradual and regular changes of species went on during those periods, though all its intermediate links and steps are lost to us, and only the extreme terms are preserved.

Non-fossiliferous intervals in one and the same formation.

We have one striking proof of this, in the fact, perfectly familiar to geologists, that in many formations we frequently encounter a thin layer of their characteristic fossils, upon which succeeds a large, and sometimes an enormous, thickness of the same deposit, *wholly destitute of organic remains;* after which again occurs another thin layer *full of them;* and this sometimes *repeated* more than once *in the same formation;* a distinct proof, therefore, that, while these beds, destitute of *all indications* of animal life, were being deposited (which must often have been a period of great length), animal life was still really going on in full intensity and variety, though from local causes no specimens of it were imbedded; yet its continued existence was evinced again when the upper fossiliferous bed came to be deposited.

Recapitulation.

To recapitulate:—The argument from the known to the unknown is clearly this: in one instance we find two different epochs, at which species or even genera exhibit a wide difference: of the interval between these we have, however, continuous evidence showing that, during this vast period, species have gone on changing by insensible gradations, until, taking its two extreme points, they exhibit that wide difference alluded to. Again, in another instance we find two different epochs, at which species or even genera exhibit a like wide difference. Of the interval we know nothing: we have either no intermediate beds, or an azoic mass. The obvious inference from analogy is, that that interval was probably as long, and was marked by as gradual changes, as the former, though circumstances have prevented their being exhibited to us.

Break in succession of species only shows long interval of time

If, then, we find, a bed containing certain species, and then superimposed on it another containing forms, not only specifically, but even generically, or still more widely, different, instead of a real hiatus, an interruption, a destruction, and a sudden reproduction of life, the fair inference would be the occurrence of an indefinitely long interval of ages, during which, indeed, no fossiliferous deposits took place at that locality, but during which the slow progressive change of species went on, until whole genera were different, and then a deposit took place in which some of these latest remains were imbedded. *The wide organic difference between two contiguous beds would only mark the longer interval of time between their deposition.*

In confirmation of the ideas thus suggested, I have great satisfaction in citing the testimony of two very distinguished men, each delivered from the chair of the Geological Society. The first I will quote is a single passage from the anniversary address of Mr. Horner, who, amid a variety of other able remarks bearing on the present subject, observes, "By whatever names we designate geological periods, there appear to exist no clearly defined boundaries between them in reference to the whole earth. Such a marked line may be seen in particular localities, but every year's experience, and our more intimate acquaintance with the phenomena exhibited in different countries, and with the distri-

Views of Mr. Horne

bution, structure, and habits of animals and vegetables, teach us that there is a blending, a gradual and insensible passage from the lowest to the highest sedimentary strata, particularly in respect of fossil remains. The terms we employ to designate formations can only be considered as expressing the general predominance of certain characters, to be used provisionally, as a convenient mode of classifying the facts we collect, whilst that knowledge is accumulating, which in after ages will unravel the complicated changes that belong to the successive periods into which the history of the structure of the whole earth may be divided." *

Views of Professor E. Forbes.

The second opinion which I have to quote, is that of the late Professor E. Forbes, who says, "I am one of those who hold, *a priori*, that all gaps are local, and that there is a probability, at some future time, of our discovering gradually, somewhere on the earth's crust, evidence of the missing links. All our experience and knowledge, theoretical and practical, warrant the affirmation that, at every known stage of geological time, there were sea

* Address to the Geological Society, 1847, by Leonard Horner, Esq., President, p. 22.

and land. Even those who believe in a primeval azoic period will hardly sanction the supposition that there has been any repetition of azoic epochs since the first life-bearing era commenced. And if so, and if there were always sea and land since the commencement of the first fossiliferous formation, we are warranted in assuming that both earth and water had their floras and their faunas."

"All geological experience goes to show that, whenever you have a perfect sequence of formations accumulating in the same medium, air or water, as the case may be, there is, if not a continuance of the same specific types, a graduated succession and interlacement of types, and of the facies of life-assemblages; even as on the present surface of the earth, the faunas and floras of proximate provinces intermingle more or less specifically; or, if physical barriers prevent the diffusion of species, assume, more or less, one general facies. This passage by aspect and type of one stage in time into another, is but scantily indicated at present in the uppermost manifestations of palæozoic life, and the lowermost of the mesozoic. The missing links will sooner or later reward the diligence of the geological explorer." *

His theory of polarity.

The author, however, conceives that, notwithstanding this general unity, there are features in the distribution of organic existence in time, which seem to indicate some real law not as yet recognised; especially in the instance referred to of the slighter connexion in sequence between the mesozoic and palæozoic periods; and he proceeds to suggest an explanation by applying a new theoretical idea of the convergency, as it were, in time, of certain groups of forms towards a point of a greater intensity, which he designates by the term "*Polarity;*" a theory which it would be impossible here to discuss, but which, from its important bearings, as well as the deep interest it carries with it as being the last speculation he lived to propose, will doubtless command the closest attention of philosophical geologists.

Continuity in the tertiary period.

Speaking of the *tertiary* formations, Sir C. Lyell observes, "There are usually so many species in common to the groups which stand next in succession

* Proceedings of Geological Society, Address, 1854, No. 38. p. 78.

as to show that there is no great chasm, no signs of a crisis, when one class of organic beings was annihilated to give place suddenly to another. This analogy, therefore, derived from a period of the earth's history which can best be compared with the present state of things, and more thoroughly investigated than any other, leads us to the conclusion that the *extinction* and CREATION of species has been, and is, the result of a slow and gradual change in the organic world." *

Again, he argues at length, from the actual causes which determine the conditions of successive deposits that, assuming, as we thus must do, the "fluctuations in the animate world to be brought about by the slow and successive *removal and creation of species*," yet from the local nature of the formations we cannot expect to find conditions such as shall enable us to trace the "*gradual passage* from one state of organic life to another."†

Lamarck's view of breaks.

Lamarck, indeed, held that there may be some gaps in the series greater than we can attribute to mere want of evidence, or hope to see filled up by

* Principles of Geology, p. 179. 8th Edition. † Ib. p. 184.

future discoveries; yet he conceived that the difficulty might be obviated from the consideration of the counteracting influence of a variety of external causes, which are perpetually interfering with the regular order of succession. If these interfering causes did not exist, we might expect an exact continuity of forms; but by these immensely varied agencies of external and local conditions, the progress of some races may be retarded, and that of others accelerated; so that at length wide breaks of continuity may necessarily appear after a long lapse of time.

In many cases too, it must be recollected, that the apparent interruption is confined to certain classes of animals only, and not to others; chiefly to the higher forms; while in the lower during the same periods, no such interruption occurs, some of them being persistent through many epochs.

Appeal to general principle of uniformity and continuity.

But supposing the existence of such apparent gaps or breaches of continuity granted, and that we failed to explain them by any such theoretical suggestions, although we may not yet have hit upon the true explanation or traced *the particular law* in this case, we are sure that *some law* is really involved even in a seeming infraction of a regular series, and cannot doubt that future discovery will ultimately disprove or explain the apparent anomaly. So strong is the inductive assurance of this, that we may safely allow any such apparent exceptions to await their solution without in the least influencing our opinion of the soundness of the broad principle of the continuity of physical causes: a principle of that truly philosophical character which no exception in detail can subvert, or render, in some form, inapplicable or unfruitful.

No real interruption.

No inductive inquirer can bring himself to believe in the existence of any *real hiatus* in the continuity of physical laws in past eras more than in the existing order of things; or to imagine that changes, however seemingly abrupt, can have been brought about except by the gradual agency of some regular causes. On such principles the whole superstructure of rational geology entirely reposes; to deny them in any instance would be to endanger all science.

There is no force in such a merely negative argument; we cannot doubt that the seemingly disjointed portions of the chain must be *really* as much connected as in the more recent instances, where we can see its continuity, and that some future research will as fully and surely close up the apparent breach, as former discoveries have done others once quite as wide.

No sudden or inexplicable agency necessary.

Thus enough has probably been said to show how completely fallacious is the inference that in such cases as those referred to, because we find an apparent interruption in the observed series of organic remains, therefore we are to conclude a real interruption in the order and continuity of organic existence. And still further from all sound reasoning or rational analogy must be the inference that, when we find in a superior bed, animal remains seemingly disconnected with those in an inferior, the actual origination of those distinct species was, therefore, in any way of a sudden or peculiar kind disconnected with the preceding order of things, or the orderly progress of natural causes.

Continued permanence of physical laws in the inorganic world.

Throughout all the immense periods of the primeval earth in its manifold mutations, the researches of the geologist present to our contemplation *two broad facts* in most remarkable juxtaposition; the invariable *constancy* of the nature and laws of *inorganic* matter and of the forces acting on it, under all the revolutions affecting it, on the one hand, coupled, on the other, with the perpetual indications of *change and fluctuation* in the forms and functions of *organised existence;* and the question arises—Can this fluctuation and change be otherwise than the result of equally invariable though unknown laws, applying to the organic world?

Thus in the *inorganic* world we trace the same slow and gradual elevations and depressions of continents which we actually witness going on at present. The same results of earthquakes and landslips, the action of volcanoes and glaciers, of submarine currents, oceanic and fluviatile deposits, irruptions of water over depressed lands, drainage of lakes, and a multitude of like events, all happening in obedience to the same identical mechanical and hydrostatical laws in the remotest abysses of past time, as they do at this day; the same influences of the seasons, and even variations in them, stamped in the concentric interior rings of fossil trees.

We find the evidence (so beautifully illustrated by the researches of Sir C. Lyell), even back to some of the earliest epochs, of the existence of the same atmospheric conditions; the rain-drops imprinted on the mud; even the obliquity of its descent, indicative of the force and direction of the wind. The very existence of such drops implying the same action of atmospheric electricity and the laws of cohesion; the power of the sun to dry up the mud (implying heat conveyed, as now, in the rays of light), thus preserving the impressions of the footsteps of animals on the wet surface left bare by the sea during a short interval, to be covered over by a fresh light deposit by the returning tide, whose recurrence evinces by consequence the same laws of cosmical gravitation.

Continued changes in the organic world.

But the unchangeableness of mechanical laws is always found under continual changes of outward conditions, corresponding to which we trace through the series of organised life perpetual and unceasing variations of forms and species, yet carried on with such slowness, that we only perceive it by comparison at immense intervals of time. In like manner, of organised life we find some of the conditions equally unchanged; the animals and plants of those remote epochs, even like those now existing, subject to the same general physiological laws of respiration and circulation, digestion and nutrition, locomotion and instincts; their eyes and ears adapted to the same optical and acoustical conditions; their reproduction generally regulated by the same laws; and *during comparatively limited periods, and identity of external condition, the same permanence of species.* But amid all these indications of uniformity, when we come to compare the state of things after immensely long intervals, we find the nature of whole tribes has been undergoing metamorphoses; not arbitrary or heterogeneous in their character, but often repeated in regular correspondence with other inorganic changes, according to some uniform plan whose law is not as yet made out; but in all their changes corresponding strictly to the modifications of one common primitive type according to recondite laws of analogy.

Introduction of new species regular, not casual.

But however little we know of the laws or causes of these changes, one thing is perfectly clear, *the introduction of new species was a regular, not a casual phenomenon;* it was not one preceding or transcending the order of nature; it was a case occurring in the midst of ordinary operations going on in accordance with ordinary causes. The introduction of a new species (however marvellous and inexplicable some theorists may choose to imagine it) is not a solitary occurrence. It reappears constantly in the lapse of geological ages. It recurs regularly in connexion with those changes which determined the peculiar characters we now distinguish in different formations. It is part of a series. But a series indicates a principle of regularity and law, as much in organic as in inorganic changes. The event is part of a regularly ordained mechanism of the evolution of the existing world out of former conditions, and as much subject to regular laws as any changes now taking place. If the series be regular, its subordinate links must *each* be so; the part cannot be less subject to law than the whole. That new species should be subject to exactly the same general laws of structure, growth, nutrition, and all other functions of organic life, and yet in the single instance of their mode of birth or origin should constitute exceptions to all physical law, is an incongruity so preposterous that no inductive mind can for a moment entertain it. It must have been as truly subject to pre-arranged laws as any case of ordinary reproduction.

Due to regular natural causes.

And since in any conjecture as to the nature of the causes acting to produce these changes, we must admit that *long duration of time* necessarily enters as an essential element, it is obvious that we can in no way form legitimate inferences respecting those causes from any mere observation of natural operations *which do not require time* for their evolution, or conclude against such changes having occurred, even to a great extent, in those immensely long periods, because we do not see them occurring in a short time under our own eyes, in the brief and momentary periods to which our observations extend. Nor in this view is anything implied adverse to the strict application of the truly philosophical principle of arguing solely from real physical causes for the explanation of geological phenomena. Sir C. Lyell expressly includes *lapse of time* as an element among the conditions which he lays down in that grand maxim, worthy to have occurred in the "Novum Organon"—"*When we are unable to explain the movements of past changes, it is always more probable that the difficulty arises from our ignorance of all the existing agents, or all their possible effects in an indefinite lapse of time, than that*

Influence of time.

Appeal to real causes in conjunction with effects of time.

some cause was formerly in operation which has ceased to act."

Conclusion. In a word, if we acknowledge the right mode of investigating the organic phenomena attending the gradual formation of the earth's crust, like all others to be solely that which proceeds by the analogy of real physical causes, carried on through countless myriads of ages, not by the agency of imaginary convulsive paroxysms, then by the same rule, the same principles ought to apply in regard to those more obscure changes of an *organic* kind continually going on, whose nature, indeed, is less understood, but which, therefore, form not less an integrant part in the prescribed and beautifully adjusted economy of nature.

§ II.—THE EVIDENCE DERIVED FROM PHYSIOLOGY.

Physiology essential to the inquiry.

GEOLOGY is essentially dependent on physiology; hence any argument derived from the former science as bearing on the evidence of "creation," must be in some degree an application of the latter; as, indeed, is manifest throughout the foregoing remarks. But some questions are involved in the present inquiry, which depend on a more particular reference to points of *pure physiology*, and to these the present section relates.

Researches of Cuvier.

From the researches of Cuvier, the whole science of comparative anatomy received a vast, and at the time unimagined, extension in its application to the organic remains of the ancient earth (first systematically carried out by Von Buch), and the recognition of extinct species, allied to existing forms, in what were hitherto imagined to be either relics of legendary monsters and antediluvian giants, or else mere "lusus Naturæ" created by her "plastic powers."

The intellectual character of Cuvier was equally marked by high powers of generalisation and by a dislike of theorising, or indulging in speculations as to the causes of the changes of form observed. He inclined, moreover, very strongly to the idea of investigating organised structures on the principle which he termed "conditions of existence," or what has been since called "teleology."

Views of Geoffroy.

Meanwhile, the rising school of Geoffroy St. Hilaire, in proposing the principle of "unity of composition" as that on which alone a philosophical investigation of organised structures ought to be built, was strongly opposed to the followers of Cuvier, especially in reference to the unphilosophical use of the appeal to final causes, in accordance with what has been already explained.* While, in regard to the relations of species, in some respects they pushed their physiological speculations into the regions of conjecture beyond the boundaries within which demonstrative evidence had as yet been applied.

* See above, Essay I. §§ ii. and v.

In the sharp discussions which these questions underwent in the Academy of Sciences, it was not surprising that under the high influence of Cuvier's name a considerable body should have stood out antagonists to the doctrines of the newer and transcendental school, and stout maintainers of the *safer* dogmas, both with respect to the "teleological" principle, and in opposition to the novel theory of the relations and modifications of species.

Principle of unity of composition.

If we look at the former question in a purely scientific point of view, it amounts to the inquiry, whether in the actual organisation of animals the "governing principle," or general law, is to be regarded as that of an archetype or common plan to some modification of which every observed form may be reduced,—or whether we should rather look to the *conditions* under which we suppose each animal destined to exist, and interpret the different structures in their imagined relations to that end. Professor Owen * has justly observed that the two principles of "unity of plan" and "final causes," are "wrongly regarded as antithetical;" and on

* On Limbs, p. 84.

general grounds, it must be apparent that these two principles can hardly, with propriety, be put in opposition to each other, or even be classed together; the latter is, in its nature, a more particular and restricted kind of *practical* view of the matter, doubtless of some value in particular cases; while the former is of a comprehensive speculative character, fitted to form the foundation of a philosophical system, which the other never can be, as Bacon has so forcibly pointed out.*

Narrow views from teleology.

If more particular arguments were wanted, Professor Owen has shown precisely from instances that the mere investigation of the *uses* of organs continually finds a check in the observation of many cases where organs are introduced whose function or purpose is not fulfilled; and the more anatomical investigation has extended its bounds, the more clearly have such proofs been displayed, evincing that this principle is an insufficient guide.

Instances.

Thus, the complex form of a limb, as to number and relative position of the bones, required by the law of conformity to the type, is strictly preserved

* See above, Essay I. § v. p. 139. And the whole passage in De Augm. lib. iii. c. 4.

in cases where it is not needed; as is seen in comparing the expanded human hand, where every finger is essential, with the "trowel" of the mole, the paddle of the whale, or "hoof" of the elephant; where every bone is equally present, but all enclosed in one case. So, again, the abortive teeth of the young whale are of no use except to prove its relationship with terrestrial mammalia. Unity of plan is adhered to in other cases where only one or two parts are developed, and the rest are merely rudimentary, or even altogether deficient, but no new part or formation is added. Nothing is made in vain if it be only made to preserve unity of system. The view of design has been contracted by the adoption of the false analogy of machines, in which unity of plan is not an object. The attainment of an end by apparently circuitous means for the sake of obedience to the law of unity is, in fact, the highest indication of design; special adaptation is but a secondary branch of such evidence; and it is only the more striking when brought about in conformity with this higher and governing principle of all animated nature.

With regard to the principle of "unity of com-

Enlarged ideas of unity of composition.

position" itself, some general and vague notions of the uniformity of plan pervading animal structures had been thrown out even by Aristotle; and in modern times, Bacon had recommended an inquiry into the *causes* of the diversity of organised forms, and Leibnitz and Newton had both hinted at the idea of a common plan pervading animal structures. And these rudimentary conceptions were, perhaps, first developed in a more systematic, yet hypothetical form in the speculations of Goëthe and Oken, and more extensively in those of Geoffroy *, in whose school it had been fully recognised as the most material point of comparative anatomy to establish the analogy of the several functional parts in different species on what was termed the doctrine of "homologies."

Numerous and striking instances had been long since pointed out, which show under evident *dissimilarity*, the extent to which real *analogy* is preserved. In many instances the fully developed organ in one animal structure is only found in a rudimentary condition in another; so much so, some

* Principes de Philosophie Zoologique, 1830.

times, as to require the minutest examination to detect it, yet, still strictly *homologous,* or preserving the same *relation* to the general structure; and so strong was the conviction of this general law, that many physiologists did not hesitate to speak positively of such analogies on the strength of conjecture, where actual examination had not yet detected them, thus laying themselves open to the attacks of more matter-of-fact inquirers.

Theoretical speculations often useful.

Some of these views, especially those of Oken, were, perhaps, of too metaphysical a cast to be usefully recognised by physiologists. Yet, they at least fulfilled the important purpose of supplying hints and presumptive conjectures for a more exact induction to work upon; and whatever, on various grounds, may have been the prejudices against the views of the transcendental school, physiologists are now beginning to pay them the homage of carrying out and establishing on demonstrative evidence (at least in regard to this great principle) the ideas which they suggested.

Analogies of vetebrate forms.

To proceed to a more particular view. Goëthe and Oken had thrown out the singular analogical idea, that the bones of the skull are all vertebræ;

and others had imagined some vague resemblances between limbs and ribs, as connected with the vertebral structure. But it was reserved for Owen to give the full elucidation and establishment of these views, and to supply a detailed demonstration of a principle so essential in the theory of unity of composition of animal forms.

Owen's researches on the vertebrate form.

By a close anatomical investigation of the true nature of a vertebra, he confirms the idea of the vertebral character of the bones of the cranium, and includes the whole structure of the vertebral column in a single analogy; and thus traces the limbs to the development of certain appendages which he has shown to belong to all vertebræ; but in different cases more or less detached and displaced from those vertebræ, and developed in different degrees, in adaptation to the respective forms and functions.

Thus the whole skeleton is referrible to one simple scheme or archetype, most resembling the *fish* form; to analogy with which all the most varied modifications may still be traced.*

* These views are most luminously set forth in detail in Professor Owen's small volume "On the Nature of Limbs." London, 1849.

In the Appendix I have inserted a more copious abstract, which has received the benefit of Professor Owen's own revision and remarks.

The idea of "unity of composition" as at first proposed by Geoffroy, like many other ideas struck out by great master minds, was a kind of philosophical prophecy: he did not himself carry out the investigation of it in all its details by demonstrative proofs; and even in some instances it has been pointed out that he fell into mistakes in particular points of its application. The research remained to be fully followed out by others: and as each of the great divisions of animal life — the Vertebrata, Articulata, Mollusca, and Radiata—had been shown to have *separately a plan and an archetype of its own*, the question which then arose was — Can these four great rudimentary plans be shown to have a yet more comprehensive relation? Can they be included under any *one common and yet more elevated generalisation?*

Generalisation of unity of composition.

According to the system of comparing *structure* alone, it was impossible to establish such a point; a different method was necessary: and Von Bär was the first to suggest the principle of studying and comparing, not merely the *adult structure*, but the earlier process of *development* of each form: and in following out this line of research he was able to

From comparing fœtal structure.

indicate characters assumed at successive early stages by the different organs; and thus to trace the relations of species and classes in a way which the examination of the mature structure alone would not have admitted; and he thus showed that, though the *common plans* of the *adult* forms of the great classes are *dissimilar*, yet in their respective developments there is in each *a period* during which an *exact community of plan* prevails: beyond that stage they diverge according to laws peculiar to each.

It has been in carrying out this "developmental method" of comparison that the labours of subsequent inquirers in this field have succeeded in the full establishment, in detailed anatomical examination, of the great idea of unity of composition: and what was at first little more than a philosophical romance, has in their hands risen to the rank of a demonstrated science.

Analogy of vertebrate and invertebrate classes.

The investigations of Professor Owen forcibly elucidate, not only the correspondence traceable between the perfect organs and functions in different species, but also the relations existing between the *permanent* organisation of the lower classes of animals, and the *transitory* embryonic steps through which the higher pass, and evince the extent to which the resemblances expressed by the term "Unity of Organisation" may be traced between the higher and lower organised animals; and that it bears an inverse ratio to their approximation to maturity. "All animals," he observes, "resemble each other at the earliest period of their development;" and he traces out with precision the characteristics which mark each stage of development as compared with those which permanently belong to different inferior classes, from "the monad," with which alone "the potential germ of the mammal can be compared," up to the vertebrated form in its different modifications.*

In this way the Annulosa having been analysed to a common original rudiment of form with the Vertebrata, by the labours of Savigny, Andouin, Milne-Edwards, and Newport; so the Mollusca and Radiata, more recently examined, have been reduced to a similar conformity to the same principle, especially by the labours of Mr. Huxley, who has

* Lectures on Invertebrate Animals, &c. 1843.

indicated the archetypal form belonging to them in common with the former up to a certain stage, beyond which their peculiarity of character is superinduced.*

Nature of species in general.

In advancing from these researches on *what is*, to their application to *what has been*, and with a view to the discussions which have arisen on such questions, it becomes necessary, in the first instance, to advert briefly to some considerations as to the nature and distinctions of *species* in general.

According to the distinction usually maintained in natural history, a *species* is not merely the logical subdivision of a *genus*, but implies the idea of distinctive characteristics *derived from a parent*, and the reproduction of like individuals: it involves, not only the consideration of *type*, but of *descent.*

Again, it is within the bounds of *observation of the existing order of things* to recognise the fact that these characteristics in any one species are not *absolutely* fixed, but admit of a certain and often

Varieties.

very considerable variation from one individual to

* To Mr. Huxley I am indebted for a valuable original sketch of these investigations, from which, by his permission, I am enabled to present to the reader a very copious extract in the Appendix.

another. There are, also, cases in which certain deviations from the original type are more marked, and continue to affect several or even many generations constituting "*a variety.*"

These varieties after a longer or shorter period, *in many cases* cease to be continued, and it is probable that the same external conditions, which are favourable to the original type, are less so to the variety; which is thus more easily checked in its increase, or at length extinguished. But it is a subject on which nothing is known as to the real causes which may give rise to such changes, and on which, therefore, it is clearly *unwarrantable to dogmatise*, or to reason upon such failure as if it were a *necessary* law.

Occasionally permanent.

There are also cases in which varieties have been found to continue so long, and to maintain so completely distinct a character, that it has become difficult, if not impossible, to determine whether they do not constitute a *sub-species.* And so far as any speculation can be carried on a subject so little understood, it would seem most probable that wherever such permanency has been attained, there is always some peculiarity in external conditions which has been the determining cause for the perpetuation of such variety, just as some other external conditions of an opposite kind are in other cases unfavourable, and cause its declension or extinction. It is therefore a fair inference that, if the favourable conditions were continued, and the variety were locally isolated from the rest of the species, it would become a permanent type or species.

Argument for permanence of species.

Within certain limits we observe species *fixed* at the present day: we have, in some instances, proofs from historical monuments and preserved remains that they have not altered within very high limits of antiquity. Some writers refer, for example, to the mummies of Egypt, reaching back to an interval of 3000 years (or, indeed, as much longer as their chronology may dictate); but, in fact, we can go much higher, since we have undoubted evidence of *some* existing species having remained permanent during the countless ages since the tertiary deposits up to the present time.

Permanence of certain species accompanied by changes in others.

This, however, proves little, since the point to be explained is, that associated with these, are found other species closely allied to, but different from, those now existing, and since introduced; and the question is as to the relation between the existing species and its ancient analogue, or generally between any species existing in one formation or period, and the one analogous, or nearest allied to it, in the next earlier formation in which it had no place.

Difficulty of marking specific distinctions.

Again, it is the fact that there are at present numerous instances in which naturalists are at issue as to what constitutes a *variety* or a *species:* and the very difficulty of determining whether any given character is or is not permanent, is alone sufficient to show how little we ought to press any such principle as the basis of a universal conclusion — still less as partaking in the character of a great and necessary law of nature, on which we can satisfactorily reason from the present to the past, or infer the perpetuity of distinctions so little settled.

The distinctions admitted as those of *species* are indeed in some cases much more minute than in other instances, those considered to mark only *varieties.* The American butterfly the (Vanesso Atalanta) is held to be a distinct species, yet its characteristic is only a single spot on the wing. Nevertheless, the entire black skin of the Negro and the white of the European mark only varieties.

In a report formally put forth by the British Association, it is expressly set down as a DESIDERATUM to discover some sure mode of *distinguishing real species from local varieties.* "One school," the reporter observes, "attribute all specific distinctions to the influence of external agents; another regard the most trivial circumstances as fixed;" but he recommends a just mean as the preferable view.*

Professor Pictet also enlarges on the difficulty of distinguishing species, and gives instances of such ambiguity of character.†

On this subject I will again cite the authority of Dr. Carpenter:—"The uncertainty of the limits of species is daily becoming more and more evident, and every naturalist is aware that a very large number of races are usually considered as having a distinct origin, when they are nothing more than permanent varieties of a common stock."‡ And he then goes on to point out the course which the true naturalist must take in the endeavour to define them

* British Association Report, 1844, p. 219., On Ornithology, by the late H. E. Strickland, Esq.
† Geological Quarterly Journal, No. v. p. 48.
‡ Principles of Physiology, Art. 542.

more precisely by attending, not merely to form and structure, but to "the *whole* natural history of a reputed species," to ascertain and limit its real character.

But if such "permanent varieties" present characters so constant, the real bearing of the question is evident, Why may not other or all "reputed species" have once originated in the same way?

Increase in number of known species.

And here, indeed, another somewhat material question arises: as discoveries and explorations extend and increase, so must the *number* of known species eventually extend and increase in an almost incalculable ratio; and the inference clearly is, that as new species are thus continually being *inserted* between other allied species already known, it is evident that the specific differences between each must tend to diminish continually, and *all species* tend to be connected by more and more close affinities.

Real number of species infinite.

It has been stated on good authority that by the recent progress of research the number of *known* species of plants and animals has been *doubled* in some classes, and *quadrupled* in others, within the memory of persons now living; and considering the number of *new species* constantly reported in every fresh exploration, in every successive number of every journal devoted to natural science, we cannot but suppose the increase to continue at least at an equally rapid rate.

And differences evanescent.

But while the number of species thus tends to become infinitely great, the extreme difference between man (let us suppose) at one end, and a zoophyte at the other end of the scale, is constant and *finite;* hence the average difference between any two species tends to become *infinitely small.* Multiplied by the number of species, it must still be equal to a *finite* quantity; and the product being *finite,* if the first factor be *infinity,* the second must be *zero.*

Yet sufficient practical distinction.

The close approximation in character between many allied species has led some philosophers to speculate on the real difficulty of any absolute and philosophic distinction between them; for all *practical purposes,* indeed, and in reference to the *existing state* of knowledge, and any state which it may be likely for a long time to assume, there is in general sufficient ground of distinction.

The alarm felt lest the power of making specific

distinctions should be thus done away, and with it all substantial science of classification should disappear, is seen to be groundless when we observe that it is on all hands allowed that species are likely to be as strictly permanent as at present for many thousands, perhaps millions of years to come, *provided the external conditions remain the same.* When the distribution of continents and oceans, the elevations of land, the direction of currents, and the like circumstances, shall have undergone a great and notable change, influencing the climate and productions of existing lands, and even presenting new regions for the diffusion of life, we might then well expect that forms of existing species might be lost, and that such an entire change of species, and eventually even of whole genera, might gradually take place as would fully exemplify, and account for, the observed changes in ancient formations.

Species permanent for the present era.

Much discussion (as is well known) has arisen on the question whether the *different races of men* are *varieties* of one species, or *distinct species:* and it seems to be at present the prevailing opinion that they are *varieties* merely.* But the question *how,*

Races of men, whether species or varieties.

* For example, Dr. Pickering, after an extensive investigation, con-

by what steps or processes, did such large and fundamental differences arise? entails more important consequences than many in their zeal to maintain a single origin, seem to perceive. It is clear that these differences are fully as great as those which in many other cases are allowed to constitute distinct species.

Difficulties of the question.

If in the case of man they have occurred as transitional varieties, how comes it that they have become so inveterately permanent? And if those changes have all occurred within the lapse of a few thousand years of the received chronology, it cannot with any reason be denied that similar changes might occur among inferior animals, and become just as permanent. And if so, changes to an indefinitely greater extent might occur in indefinite lapse of time. If these changes take place by the gradual operation of natural causes, it would be preposterous to deny the possibility of equal or greater changes by equally natural causes in other species in equal

---

cludes that there are either *eleven* species, or only *one*, of the human race. But upon further examination he decides in favour of *one*; and thinks the original seat of man was in Africa. (The Races of Men, &c. by C. Pickering, M.D. Lond. 1850.)

Others have assigned *six*, or other numbers of species: the author of the "Vestiges" supposes *two* local centres necessary. (P. 220. 6th Ed.)

or greater periods of time. The advocates of the fixity of species would argue that the single spot on a butterfly's wing, which constitutes a species, never has changed, and never can change, without a miracle; and yet the vast differences between a European and a Negro or Australian are mere modifications of one parent stock by natural causes in the lapse of a few thousand years!

The peculiar characters of the Negro race are recorded as prominently marked, as at present, in the ancient Egyptian paintings, which may go back 3000 years or more.* Here, then, is a *variety* which has been *permanent* for at least that long period; a period, too, which has been expressly relied on by many to prove the permanence of *species* by appeal to these very monuments. And then we have to ask, How long must it have taken, at this rate of imperceptible progress, to have been developed out of the original stock?

Varieties of the dog.

Another instance has been much dwelt upon, the so called "varieties" of the dog, presumed to be derived from a common stock; but how long since,

* See Kenrick on Primeval History, p. 20.

is undetermined. Yet in these varieties (in which even the form of the cranium greatly differs) it would be difficult to deny that the distinctive characters are permanent, at least under the continuance of the same external conditions; and that each race, when preserved isolated under such conditions, would remain permanently distinct.

Hybrids.

Much stress has also been laid by some on the asserted sterility of hybrids; though, in truth, it affects very little the general question; while its very limited evidence dependent only on a few isolated facts, occurring in a state of domestication, is utterly insufficient for the foundation of any general law. The cases commonly referred to should be regarded by an unprejudiced mind as probably exceptional, under peculiar conditions, and not to be dogmatised upon, as involving any real and necessary law of organised existence. As there are limits beyond which union will not take place, so within these there may very probably be *certain limits* of still nearer affinity, *beyond which* sterility in the offspring prevails, but which have not yet been determined. The recurrence to the original type often observed, only proves that conditions are not favourable to the continuance of the variety.* And of the very positive assertions so liberally made in these and the like cases, it is to be observed that they are, at best, merely *empirical* conclusions, wholly unsupported by any wide analogies, or explained by any known causes which can confer on them the character of real natural principles.

Conclusions of this kind empirical.

Yet the immutability of species, as something essential to their nature and inherent in it, has been upheld by a large section of naturalists—and still more strenuously by some who are not naturalists—in this country, with a degree of positiveness and even vehemence, which the mere *negative* character of the evidence could never justify, and which it would be difficult to account for, so far as any arguments of a philosophical nature may be supposed to influence the opinion.

Hypotheses of immutability of species.

It is indeed difficult to say what extent of mysticism is not connected in the minds of some with the notion of the immutability of species. Even such sober naturalists as MM. Agassiz and Gould†

* On the subject of hybrids, some important remarks will be found in Dr. Carpenter's Physiology, Art. 545.

† Principles of Zoology, p. 43.

speak of it as dependent on an "immaterial principle" essential to animal life.

Opposite views.

But in other schools, especially on the Continent, opposite views are extensively maintained, and probably gaining ground. In the case of plants more particularly, it is simply as a *question of facts*, that some eminent botanists view the matter. Thus one of the most distinguished foreign naturalists, Prof. Schleiden of Jena, after giving a variety of illustrative instances, thus sums up the state of the case:—

"We know that varieties once formed, when they have continued to vegetate under the same conditions for several generations, pass into sub-species; that is, into varieties which may be propagated with certainty by their seeds. How, then, if the same influences which have called forth an aberration from the original form of the plant, continue to act in the same way, not for centuries or tens of centuries, but for ten or a hundred thousand years, will not at last, as the variety thus becomes a sub-species, so also, this, become so permanent, that we shall and must describe it as a species." *

* "The Plant," &c. by M. S. Schleiden, M.D., Professor of Botany, Jena. Transl. Lond 1848. Lect. xi. pp. 272. 290.

Even in this country, Dr. Lindley has ventured to state not less alarming facts respecting the class Thallogens (including the seaweeds, fungi, and lichens), to the effect that "in their simplest forms all trace of series is missing," and that their species seem all convertible into each other under particular conditions.*

On these questions I have much satisfaction in referring to the opinions of Dr. Carpenter †, put forth with equal candour and freedom from prejudice, and with a union of caution and enlargement of view which eminently commends them to the convictions of the reader.

He observes ‡, "Our belief that the new beings formed by the process of reproduction always closely resemble the parent stock, is certainly founded upon a limited induction from observations made upon the higher classes of plants and animals. Reasons

* Vegetable Kingdom, 1846, p. 5.

† A number of important examples are collected by Dr. Carpenter, "Physiology," Art. 547, 548.

I do not here profess to go into details of particular instances, but many such, of transmigrations of existing species, which have been collected, are unassailable evidence as far as they go. (See for example, "Vestiges," p. 136. *et. seq.* 6th Ed.) See also a paper in the British Association Report, 1852, sectional proceedings, p. 68., by Major Munro, F.L.S., who limits transmutation to plants of the same genus.

‡ Physiology, Art. 517.

have already been given for the opinion that the same germ may assume very dissimilar forms according to the circumstances under which it is developed: and knowing, as we do, how readily the simpler classes of organised beings are affected by changes in their external conditions, it is not difficult to admit the possibility of their forms being thus greatly modified, as well as of the continued propagation of the varieties thus produced."

Origin of varieties.

As to the origin of varieties, it has been admitted by most physiologists as a general principle, that those peculiar "vital forces" (of whatever nature they may internally be) are always the main acting cause, but are subject to modifications from external causes: and hence it is the preservation or interruption of the balance between these antagonistic causes which determines the formation or the modification of the type, so as in the one case to keep it up, in the other to produce *varieties.*

Again, anatomical inquirers have chiefly confined their investigations to the normal forms of organised beings. Autenrieth of Tübingen seems to have been one of the first to dwell upon the necessity of taking into account, in comparative anatomy, not merely the *perfect* adult structure, but all the varieties of *abnormal* structure which are occasionally witnessed, and to which the absurd name of "lusus naturæ" has been given, just as organic fossils were once ascribed to the plastic powers of nature, or elevations and subsidences to mysterious convulsions. He observes, "These varieties . . . are not haphazard formations; they are the remains of structure common to all embryos; they indicate the transitions through which man and all other animals are passing from their embryonic condition to the adult." Should anything interfere with this transition, the forms persist; and this constitutes what is improperly called *a variety,* and supposed to be something deviating from the regular law. "But the laws of deformation are as regular as the laws of formation. The varieties are arrests of development; they prove the unity of organisation and of type with which Nature starts in the formation of all that lives." *

Such considerations form a necessary preliminary

* Quoted by Dr. Knox, "Great Anatomists," &c. p. 59., who also cites many instances of imperfect development, and other cases bearing on the question, p. 108.

Causes of changes in species in past ages.

to any examination of the theories of change of species in the earlier epochs of our globe. Cuvier (as it has been well expressed)* was the first to give "a history of the earth, not founded on fables, but on facts;" but it was incorrectly called a "theory of the earth;" it was really only a "history," whereon to build a "theory."

On the theory — the philosophy of that history — it was reserved for others to speculate. The laws which regulated the succession of living forms in the different epochs of the earth's existence, and those higher generalisations which might tend to indicate the physical causes of those changes, were the objects of inquiry to a considerable section of continental physiologists. When philosophers began to speculate on the possible causes of changes of species in the ancient world, it could hardly be otherwise, in an inquiry of so wide and novel a character, than that the several speculations should be but imperfect in detail, and should be rather guesses at and approaches towards the truth, and should, even in the opinion of adherents to the

* Knox, p. 25.

general principle of some regular process of evolution, be seen to require revisions and retractations in minor particulars.

Cuvier, while he professed a rejection of all hypothetical speculation, appears to have been strongly prepossessed with one hypothesis—that of the essential and eternal *immutability of species.* At any rate, under the sanction of his name it has been since maintained by many of his followers with a degree of positiveness not easy to account for on any merely philosophical grounds.*

Theory of transmutation.

But an opposite opinion began to be taken up by those of a different school, intimately connected with the more speculative and transcendental views already alluded to.

Connected with the principle of unity of composition.

When Geoffroy and his coadjutors were engaged in upholding the unity of composition of all animated structures, it was regarded by them as a natural consequence that, as all the details which mark different species were the mere modifications of

* Dr. Knox, however, the personal friend of Cuvier, states that, at least latterly, he was much inclined to modify his opinions on this point, and quotes one passage in which he says, "Nous ne croyons pas même à la possibilité d' une apparition successive des formes diverses." Great Anat. &c. p. 44., also p. 27. The expression, however, is perhaps ambiguous.

greater or less development of the different parts of the primary plan common to them all, so those developed parts had no essential or permanent place in the nature of the species. Hence, naturally arose the further idea of the possible migration or transition from one species to another, or rather from one species to a *new* modification.

Thus the theory of "unity of composition" was, in the minds of many, closely allied to that of "transmutation," which seemed to be a sort of natural sequel to it, and was, among a large school of the continental naturalists, associated with the advancing prospects thus held out to their view of the system of nature, and the attainment of a more transcendental theory of her operations.

Applied to past changes.

From the principle of unity of composition as applied to the existing animal world, it was by a natural extension that this school of physiologists were led to infer that there had been an equally close analogy preserved in *time*, and that there had been a continuous succession of the several species of the animal world, of which only a few detached fragments are preserved to us, as disclosed by geological research: a succession, as they contended, which took place in the natural course of reproduction by continual small deviations from a primitive type, according to the influence of external conditions which varied from one epoch to another, yet were determined alike by fixed and regulated laws, from the present era backwards into the abysses of past time. They conceived that in none of the varied forms of organisation which we trace has there been any new fundamental principle introduced; all are but modifications of forms now existing. Thus throughout all nature, present and past, external forms are mere accidents; development of parts in excess or in defect as changing causes led to such necessity, and modifications of parts according to the functions required to be exercised under the particular conditions. The fundamental unity of principle is that which alone is permanent and invariable, and which admits of endless adaptations according to the varying conditions of existence.

Theories of development.

The speculations of Lamarck were founded, in the first instance, on observation of the fact of the near approximation of different species. He contends that the farther our researches extend the greater we find the difficulty of distinguishing species. Apparent interruptions in the continuity of the series are continually being filled up by new discoveries, and what were supposed broad lines of separation effaced. "Everything," he observes, "passes by indivisible shades into something else."

De Maillet (under the anagram of Telliamed) and Lamarck were perhaps those who followed out speculations on the transmutation of species to the greatest extent. Not contenting themselves with asserting cautiously the philosophical grounds on which the close unity of all organised structures is supported, they pursued their hypothesis into details of the most minute, and often most extraordinary, kind. Lamarck's theory, in particular, in truth presents many salient points easily open to attack, and some which are readily susceptible of being held up to ridicule. His principle, that organisation is the result of function, not function of organisation, seems, at first sight, somewhat like a mystified version of that very doctrine of final causes to which many of his school so strongly object. At any rate his theory was carried out to such unwarrantable lengths in many instances, as to discredit even the more sober and broad principles on which it was founded in the eyes of his opponents; while, as might be expected of so imaginative a speculation, increasing knowledge of facts has led to its abandonment, at least in its full extent, on the part of many who yet strenuously uphold the same broader principles. Thus Dr. Knox*, one of the most zealous supporters of the principle of transmutation in this country, speaks very slightingly of Lamarck, and regards his theory (in its extent and detail, at least) as of little weight or authority at the present day: and the necessity of great modifications in the theory is admitted by the author of the "Vestiges."†

Argument from analogy.

But in order to be applicable to the facts of the ancient earth, any process of transmutation must be imagined to extend, not only to minor features, but to a total change even of the characters which mark whole species and genera, so that entire classes and orders may in the course of indefinite progression disappear, and be replaced by others of a different kind. Lamarck endeavoured to support his theory by adducing instances actually

* Knox, p. 72.

† Vestiges, p. 143. 6th Ed.

occurring within our experience, of slight variations in species; but of more extensive changes confessedly none could be adduced from *existing* facts. Yet it was argued in such cases we must bear in mind the necessity for the introduction of another important element—*the influence of time.* Thus the allegation of actual instances of slight changes in finite times is unanswerable as far as it goes; but the absence of such evidence for greater changes in *finite* times is no argument for their non-occurrence in *indefinitely extended* duration.

Influence of time.

As in general, from connecting the conceptions of physical causes with that of immense duration of time, we obtain a very different view of the magnitude of the effects they may produce; so the continental school of transmutation regard all differences in the succession of species of past existence as rather of a chronological than of an essentially physiological kind, due to the lapse of time rather than the introduction of new physiological elements. Thus, De Blainville says that "species mark an epoch in time, not a distinction in animal nature." *

* Knox, p. 207.

Opposite views of some geologists and naturalists.

But a large section of geologists and physiologists, including many names of the highest distinction, have been strongly opposed to all speculative theories as to the probable origin of those changes in species which geology discloses. This, in many instances, has been the result simply of a rigid—perhaps over rigid—adherence to the rule of appealing to *facts* only, and allowing of no *hypothesis*; a rule which, however strictly inductive, if carried to an extreme would defeat the grand purpose of induction; and partly from imagining that, because the precise theory of development from lower to higher forms is untenable, therefore all hypotheses of the same kind are inadmissible.

Professor Pictet* denies the transmutation theory, on the ground that to produce such changes the "powers of nature" must be supposed to have been much greater in the earlier periods than we now find them to be; not allowing, apparently, for the indefinitely long time they had to work in; and, in opposition, he is disposed to maintain a series of sudden introductions of new species,

* Geological Quarterly Journal, No. V. pp. 53. 56.

which would imply powers greater still: though he afterwards seems not to consider this as quite satisfactory.

Professor Sedgwick, while he upholds the *progressive* scale of organisation in past epochs, yet earnestly repudiates the development theory, maintaining that the changes of form "mark a gradual evolution of creative power, manifested by a gradual ascent towards a higher type of being. . . . . But the elevation of the fauna of successive periods was not made by transmutation, but by creative additions."*

Real question of creation of new species.

But the question is as to the *nature* and *law* of those additions, and in what particular *sense* or *manner* they were "creative;" whether they were not made according to some determinate law, and in some fixed relation to those which preceded them or were most nearly allied to them. The question is one demanding calm and patient philosophical analysis, and will be the more fitly and worthily discussed the more expanded and generalised the views adopted of geological and cos-

* See "Studies of Cambridge," 5th Ed. Introduction, pp. 44. 154. 216.

mogonical order and progress, the more free the inquirer may be from bias and prepossession of other kinds.

Meaning of the term.

The term "creation" indeed, especially as respects new species, seems now, by common consent, to be adopted among geologists as a mere *term of convenience*, to signify simply the fact of origination of a particular form of animal or vegetable life, without implying anything as to the *precise mode* of such origination — as simply involving the assertion that a period can be assigned at which that species appears, and before which we have no evidence of its appearance. In this sense there can be no objection to its use, but it should be carefully guarded against possible misapplication.

Dr. Knox observes, "It was the opinion of Geoffroy that there never had been but one creation: this (he adds) is also my own opinion. I believe all animals to be descended from primitive forms of life forming an integral part of the globe itself, and that the successive varieties of animals and plants which the dissection of the strata of the earth clearly sets forth are due to the occurrence of geological epochs, of the power of which we cannot form any true conception."* This would seem like a leaning to the convulsionist theory, were it not more clearly explained in the following passage:—

"We know not, then, the causes of the specific and generic differences in animals, nor why such differences continue fixed for a period—the historic period, for example; they depend, no doubt, on secondary laws, which some future Newton may discover."†

The subject of Professor Owen's investigation, before referred to, acquires a higher interest when viewed in reference to the question of the progression of animal forms in past epochs of our globe, and to this view he has pointed‡ as appearing to assign a certain conformity in the order of ancient existence with that of development of the archetype, as indicated by these anatomical researches. Professor Owen, however, is specially desirous to be understood as applying his conclusions solely to the *order* and *law* of succession, without any attempt to assign a *cause or to trace its origin*.

Change of forms by

But, though anxiously disclaiming the charge of

* Knox, p. 109. † Ibid. ‡ On Limbs, p. 86.

being supposed in the slightest degree to support the development theory, he nevertheless expresses himself in the most guarded and strictly philosophical language in reference to the possible modes of explaining past changes of species:—"To what actual or secondary causes the orderly succession and progression of such organic phenomena may have been committed we are as yet ignorant."* I would only venture to add, that it must have been committed to *some* regularly ordained causes, is surely the verdict of all inductive philosophy.

regular laws.

Sir C. Lyell, though strongly opposing Lamarck's theory of development, yet freely admits, in a geological point of view, that "*If* the doctrine of changes operating in an indefinite lapse of time *be tenable*, we are at once presented with a principle of incessant change in the organic world, and no degree of dissimilarity in the plants and animals which may formerly have existed, and are found fossil, would entitle us to conclude that they may not have been the prototypes and progenitors of the species now living."†

* On Limbs, p. 86.

† Principles of Geology, p. 545.

Though the precise idea of development in one simple series, from the lowest to the highest forms, be clearly untenable, yet in what sense *some regular evolution* of successive forms may be admissible has been here discussed, and with a sufficient result, if it be only allowed to be so far conceivable as that no sudden interruptions of natural order are necessary to be resorted to in order to explain the phenomena.

Nor is the question of non-progression without a bearing on this point; since the narrower the limits of real variation of species in different epochs, the less difficult would be the application of any theory to account for them.

Local faunas and floras.

We may here properly advert to another point closely related to the question of the origin of species — the consideration of the *different faunas and floras characterising different districts of the earth.* These, as is well known, are in many instances strikingly distinct, even in regions situated near each other, and as remarkably similar in some other regions separated by intervening seas. In certain cases, too, we have a singular *parallel and corresponding series* of species in two different regions or continents, though no species are the *same* in both.

Hence naturalists have come to the conclusion that these respective systems of species must have been thus originally constituted by distinct origination peculiar to their respective localities. Thus, MM. Agassiz and Gould observe,—"There is only one way to account for the distribution of animals as we find them; namely, to suppose that they are *autochthonoi*—that is to say, originated like plants on the soil where they are found."*

Theory of specific centres.

But this view has been carried out in a more precise form by the late Professor E. Forbes, in his theory of "Specific Centres," or "Centres of Creation." This is founded on the general fact, authenticated by accumulated comparative observations (after allowing for some apparent exceptions), that numerous regions and districts of the earth are well marked out, each characterised by a flora and a fauna on the whole peculiar to itself. A given species (for example) is found to be peculiar to a

* Principles of Zoology, p. 179.

certain district; the numbers of its individuals diminish as we ascend to a particular point of geological time (subsequent, of course, to the eocene period); they also diminish at greater distances round a particular locality. Hence we recognise a single part of origin, both in space and in time, for that species: this is termed a "specific centre."

Again, in many instances, such points of origin for many distinct species occur near together, in local position, in the district characterised by those species; hence it is argued that at such points those species must have originated from an individual or pair (as the case may be), respectively the "prototype" or "protoplast" of each such species, from which that species, marked by constant specific characters, has been derived in the ordinary course of propagation. It is, however, distinctly admitted by the author, that *in what way* that prototype was formed or produced *we know not:* hence, then, and in this sense, such spots are termed "centres of creation."

And as this applies to existing species, so, the author argues, by analogy (as all organised life is part of a uniform connected system), we may fairly suppose it true of extinct faunas and floras also.*

Bearing on development.

As to the bearing of these speculations on the question of development or transmutation, it is to be observed that the only point of the theory of specific centres which bears at all on the question, is the descent of local species from a prototype of the same identical specific characters. The author expressly says,—"We have no experience of the individuals of any species being produced otherwise than from individuals of its own kind"—which is, in fact, simply *assuming* the whole question at issue. But if the *descent* of the species, unaltered through all time, were granted, the question would still remain open as to the first origin of the "protoplast," whether one or many.

It is, however, contended that distinct centres of origin are inconsistent with that view of development which traces all species to one common origin. But I am unable to see in what way it would follow

* Of this theory various details have been published by its lamented author: but I am happy to be able to present the reader with a condensed abstract of its leading principles from his own pen, with which he kindly furnished me, and which will now acquire a peculiar value and interest. (See Appendix.)

that the earliest forms of a given existing species, in a particular locality, were not modified forms of other species previously existing there; or on what grounds it could be inferred that these, in their turn, had not been modified from earlier forms; or these again, in still more remote epochs, derived from one common source.

Merely as to locality.

Thus, on the whole, the theory of "specific centres of creation" really bears very little on the question of development; it merely shows that whatever reasoning (on that theory) may be applied to the origin of species in general, must be restricted to their origination in or near *a certain locality;* in all other respects the argument, whether for or against transmutation, remains just where it stood, without reference to the theory of centres.

Still less do the facts of corresponding series of species in different continents or large districts of the earth affect the question of transmutation, since within each district such succession might have gone on separately from two parallel species now extinct. It is even possible that the more remote progenitor-species in the two districts might have been more nearly allied, the divergency increasing with each successive step.

Conclusion. Correct statement of the general facts.

On the whole, then, comparing the limited extent and purely empirical nature of our knowledge of *species* in the existing state of things, with the positive evidence of past changes, it would seem that the more correct statement of the general fact would be simply that *species (within certain limits of deviation) are permanent during very long periods, but beyond those periods a change, in some sense, occurs;* and this bears some relation to changes of external conditions. But under the same change of conditions one species may be highly susceptible of, and sensitive to, the influence of that change, while another may be insensible to it. Thus one may remain permanent, while another may undergo change, or even be exterminated. And the only question is as to the sense in which such *change of species* is to be understood;—whether individuals, naturally produced from parents, were modified by successive variations of parts, in any stage of early growth or rudimental development, until, in one or more generations, the whole species became in fact a different one;—or whether we are to

The real alternative.

believe that the *whole race* perished without re-producing itself, while, independent of it, *another new race*, or other new individuals (by whatever means), came into existence, of a nature closely allied to the last, and differing often by the slightest shades, yet *unconnected with them by descent*; whether there was a continuation and propagation of the *same principle of vitality* (in whatever germ it may be imagined to have been conveyed), or whether a *new* principle or germ originated independently of any preceding, *out of its existing inorganic elements.*

Speaking of the opinion of the formation of organised beings out of their inorganic elements, Dr. Carpenter observes — "It has been maintained by many philosophers who have regarded all matter as, in some sort, animated; and, although it has been principally urged in reference to the lowest class of beings, it does not seem possible to limit its application *if it be really valid.* . . . . Some naturalists of the present day are disposed to admit this also, and to account for the changes in the races of plants and animals which geological researches reveal."* In a

* Art. 516.

note he adds:—"Such a doctrine is impossible to refute otherwise than by an appeal to facts. No such new creations are known to us at the present time, and therefore it can only be argued from *analogy* that they ever existed. We may believe that there exists in all matter a tendency to become organised, without relinquishing the doctrine that for the maintenance of such tendency a previously existing organism is required, to collect and unite the scattered elements by the powers with which it alone is endowed. . . . . That species have in all ages of the globe maintained their present uniformity and narrow limits of variation the author is not disposed to assert, and he thinks that many facts tend to prove the relaxation at former epochs of the strictness of the laws which are at present regarded as governing their modification and reproduction."

The temperate and candid tone of these remarks offers a very satisfactory contrast to the one-sided and peremptory dogmatism we too often encounter on this subject. But it may be asked, is any "*relaxation*" in former epochs necessary to be supposed, when we simply take into account the enormous and inconceivable length of time implied in such periods? Do we require anything more than the strict observance of the very same laws of slight changes of forms accompanying corresponding changes of condition, acting through periods of incalculable length?

Legitimate conjectures from primordial unity of structure.

But while on these points we have confessedly no *positive* evidence, it is fairly open to conjecture, whether the views now universally adopted by the most enlightened physiologists of the great principle of *the unity of primordial structure* of all species may not be peculiarly suggestive with reference to such questions as those now before us.

We have already considered that grand fact on which the whole theory of unity of composition is based, the existence of a stage in the early evolution of every class and order, during which a community of form belongs to them all. At this stage there exists no difference between them, and out of this primitive common germ or rudiment any one of the more distinct specific forms might, as far as we know, be equally produced, provided the *determining causes* for that particular modification were present. Of the nature of those specific determining causes nothing whatever is at present known. It is, therefore, clearly impossible to say how far great changes of condition in external agents, or equally great changes, as slowly and gradually advancing in the more hidden internal agencies of the animal economy, might not in past ages have operated to determine successive changes in the evolution of germs originally the same into great varieties of organisation.

§ III.—GENERAL CONSIDERATIONS ARISING OUT OF THE PRECEDING EVIDENCE.

Analysis of the reasoning founded on the foregoing evidence.

THE questions to which the preceding observations refer have been the subject of much vague speculation and vehement controversy. But in the present discussion, equally desirous of avoiding the one, and deprecating the other, I wish to take a perfectly unbiassed and dispassionate view of the real tenour of the evidence; and more especially to analyse certain arguments often brought forward, and regarded as based on indisputable principles, which, nevertheless, appear to me involved in considerable doubt and fallacy. And though, in some instances, they boast the sanction of names eminent in physiology and geology, yet the question is rather one of general principles of reasoning, than of precise scientific details; and thus, without pretending to impugn their *science*, I venture to call in question their *logic*.

Argument for immutability of

In the first place, then, the belief in *the essential and inherent immutability of species*, not only in the present state of things, but as an eternal law of Nature, extending backward through all the countless ages of the ancient earth, has been upheld confessedly on the limited experience of modern observation, but thence extended by analogy in the same way (it is alleged), as *in the case of other great natural laws.*

species from permanence of natural laws.

This argument, however, appears to me altogether unfounded. Of the operation of other great natural laws, through all the series of past ages, *we have direct evidence.* The laws of gravitation, heat, light, equilibrium, and the like, present positive proof of their influence in the records preserved to us through all geological time. Whereas of the permanence of species in those past epochs, except within certain limits of particular formations, we have *no evidence* whatever: on the contrary, the *apparent* phenomena (to say the least) are all *opposed* to it; and it is the very question at issue, whether those perpetual *changes in species* which we observe, are to be considered real gradual variations in the development of organisation, or to be explained in any different way.

But difference in evidence.

Again, in regard to the mechanical laws of the

Mechanical laws proved

by principles; these only empirical.

inorganic world, *reasons* can be assigned, and calculation appealed to: whereas, in regard to *organic* life, the conclusions, such as they are, are wholly *empirical.*

The case is by no means analogous (as some seem to suppose) with that of the planetary perturbations. They would argue that, as the law of elliptic motion has held good, subject to those small deviations, through all the past existence of the system, so the permanence of species must have held good, allowing for like small deviations from type in occasional varieties. But the cases are obviously not parallel: in the former, the perturbations are all parts and consequences of the same principle and law of gravitation, perfectly understood and demonstrated by calculation to be in long periods perpetually compensated, so as to preserve the system. In the other case, the law (if it were such), is merely *empirical:* we know of no principle or reason for it. The deviations are no part of it, or consequence from it; nor can their extent be predicted. We know nothing of any causes acting, nor of any conditions which confer on it the character of necessity.

Again: the utmost extent of proof which can be

adduced in support of a permanence of species amounts to this,—that *they are permanent so long as some peculiar external conditions remain the same.* If those conditions were materially altered, no present experience will enable us to predict the result. In past epochs we know that the *conditions* were repeatedly altered; and we know that certain *changes* of species accompanied the altered conditions. Now, the stability even of the planetary system is constant only as long as the *same conditions* remain. Let them be altered in the slightest particular,—let a period, an eccentricity, or an inclination of one orbit, be changed, and the whole stability vanishes.

Dependent on certain conditions.

Within the limits of the existing period we observe a permanence of species; just as much as we find evidence of it *within like limits* in ancient periods of the earth's history: but beyond such limits of time, and the influence of certain conditions accompanying those changes, and differently affecting the several species, we have no such evidence in the one case, and direct *apparent* evidence to the contrary in the other.

Argument from conformity to experience.

But, secondly, the chief argument always is, that we have "no experience of such a thing" as change of species;—that it is an arbitrary supposition "opposed to all experience;" and the like. Now the argument of "*conformity to experience*" is beyond question the very basis of all induction, but it requires some caution in its application. And on examining the argument, as here applied, we shall find that there is a fallacy latent in the use of the expression, "we have no experience of such a thing." We may illustrate this by a familiar example.

Example from the formation of coal.

We have "no experience" of the formation of coal. Yet in past epochs we know it occurred, and it is accounted for by known and existing causes. The submergence of forests,—the accumulation of vegetable matter,—the compression of materials by superincumbent masses, whether solid or fluid,—are known natural causes, which do, or might, occur within our experience or that of history. But for the consolidation of those beds of vegetable matter, and their conversion into coal, the essential condition has been the influence of *immense duration* and vast periods of past time; and of this, undeniably, we can have "no experience."

Immense lapse of time essential to changes of species.

To apply this remark to the question of organic

changes, it is alleged *we have no experience of such a thing as a change of species;* but we have experience of the present uniformity of species *subject to slight and occasional deviations.* This is a known cause now acting. To how great an extent these successive deviations might be carried in immense periods of past time under changing external conditions, we know not.

Thus this known cause, like that of the submergence of vegetable masses, *conjoined with the influence* of incalculably vast periods of *past time,* MAY BE fully competent to give results as remote from those now *every day seen,* as the formation of coal has been, from what takes place in any submerged forest or accumulation of vegetable matter in recent times.

Thus the *mutability of species* in past epochs would not be a case impugning the doctrine of the *immutability of natural laws,* or the appeal to experience of established physical causes in accounting for past changes; because the very nature of the case *essentially involves other elements* than any now entering into our consideration. *General experience* must be understood to include more than *present, every-day experience;* and the advocates of transmutation do not assert it *absolutely*, but only in *indefinitely long periods of time.* They do not maintain that it occurs under existing conditions, but only under *great and peculiar changes of condition.* Any fair and correct statement of the case, then, must include these qualifications. *We cannot say that we have no experience of a change of species in a due length of time, and under adequate and appropriate changes of external condition.*

Correct statement of the case.

Want of experience no argument where impossible to be had.

Thirdly, closely connected with the last argument, there is another consideration equally material. Supposing it true that we have *no experience* of a particular kind of event occurring, and supposing at the same time it could be shown that, if that event *did actually occur*, we could never (from the particular nature of the case) have any *evidence* of its occurrence, then it is clear the argument from *want of experience* must fall to the ground: it would be no proof whatever of the non-occurrence of the event in question. And, further, if there were a show of reason, from analogy, that such an event were *likely* to occur, still less could the absence of experience be urged as rendering it *incredible* or inadmissible.

For example, we have "no experience," by observation, that any of the single fixed stars have planetary systems revolving round them. But it is admitted that if they had really such systems, no possible telescopic power could ever show them to us. Hence the argument, from want of observation, goes for nothing. But, farther, the stars are self-luminous and analogous to suns; and, by the same analogy, they may most probably be surrounded with planetary attendants. On the whole, then, it is a *reasonable belief* that they have such planetary systems, though all *experience* is, and for ever will be, wanting to show it.

Example. Planetary systems round the fixed stars.

Now, with respect to new species, by direct calculation, founded on a liberal assumption as to the number of species, and the probable rate of their extinction, it has been shown* that such an event as the extermination of one species, and the substitution of a new one in its place, must be *an event of so rare a character, that no noticeable instance of it could be expected to take place within the range of our observation.*

Origin of new species could not fall within our experience.

* Lyell's Principles, p. 682. 8th edit.

Applies equally to transmutation.

This argument is independent of the supposed *mode* of introduction of the new species. Hence it is to be observed, that it applies equally to the case of transmutation as to any other supposed mode of origination; and the consideration arising is clearly this—that if transmutation did really take place, *we could never expect to have any experimental evidence of it.* If it did occur, it would not fall under our notice: it is, therefore, no argument against it that we have "no experience" of it. The argument from want of evidence falls to the ground; and the hypothesis *is not, on this ground, incredible or inadmissible.*

If new species did arise, no possibility of verifying it.

But, fourthly, still farther, let it be granted that such an event as the origination of a new species were really to occur at the present day, within the possible range of observation; let us imagine that some changes were actually going on within our own times so as to develope or give rise to a *really new species,* and we ask *how could the fact ever be substantiated?* To say nothing of such facts occurring in the depths of forests and deserts, or in the recesses of the waters, where no human eye could by possibility ever detect them; let us suppose that something apparently of the kind were alleged to have been witnessed by some competent naturalist, what could it amount to more than *the existence of some new peculiarity*, which, supposing it allowed to constitute the mark of a distinct species, could, after all, only be affirmed to be *newly discovered.*

Incredulity on such a subject.

Let us even suppose it were possible for such processes *to be watched and accurately examined,* how difficult would it be for even the most skilful and unprejudiced naturalist to feel quite sure of the real nature of the case! How much more readily would any other interpretation be put upon it than that of a really new, distinct, and permanent modification; or rather, with what determined scepticism would not its reality be denied, and with what abuse and ridicule would not the unlucky observer be assailed, who should venture to assert the occurrence of such a thing as *an actual observation of the first beginning of a truly new species!*

These considerations may suffice to show how very little any reasoning from the mere *absence of evidence* will really avail in the case before us. The objection founded on it is, in fact, wholly groundless; and there does not exist the shadow of an argument to *prove* the general hypothesis of transmutation *inadmissible.*

Question between evolution of organised beings out of their inorganic elements or out of pre-existing forms.

It may, perhaps, be argued, if organic life had a beginning, there must have been some stage at which there took place *a first evolution of animal life out of inorganic elements;* and the question more precisely stated then becomes, At various, repeated, subsequent intervals, corresponding to certain epochs in the history of the globe, in order to give rise to new species, did similar fresh evolutions take place *out of inorganic matter?* or was it the case that, when certain primitive stocks had been thus constituted at first, they were also subjected to certain laws of modification of form, to come into operation under the particular combinations of external conditions which were to mark future epochs, and that so new species *were to be evolved out of the old?* The choice between two such hypothetical ideas is a perfectly legitimate subject of conjectural discussion and difference of opinion; but it is inconsistent with all inductive principles not to admit that *one or the other must be supposed.* But if the idea of a formation of organised beings out of their inorganic elements were to be preferred, still on any such hypothesis the process is imagined to be carried on through such a series of steps of gradual evolution as to differ rather in name than in essential nature, from the idea of development out of pre-existing organic forms.

If the question be ignored, it is still a *physical* question.

And farther, even if the very cautious inquirer prefer *altogether to dismiss and ignore* the consideration of the question, simply on the alleged deficiency of satisfactory evidence, still in the true inductive spirit he admits that it is *nothing more than a mere physical question* which *at present* he cannot solve. And on the same grounds he would as strenuously contend against the admission of any hypotheses derived from other considerations, of a kind incompatible with the great principles of natural order, and of a nature beyond the domain of science.

Law of succession of forms, complex.

Every advance in physiological discovery seems to point to the necessity of an entire remodelling of the very ideas of *higher* and *lower* organisation, once so much dwelt upon. The structural relations of different species, more especially of what are called the *lower* orders, are disclosed with increasing indications of the real complexity of those relations; and we are thus, palæontologically, led to a perception of the higher connections which some of the earliest *observed* forms of zoophytic* life may claim; while the true idea of *progression*, in any sense, is clearly no longer to be recognised in any single line of ascent from more simple to more complex forms, but must be sought in some new and apparently less obvious train of relation not as yet made out.

All earliest forms destroyed.

At the same time, all such speculations are deprived beyond reparation of that first essential to their completeness, a knowledge of what were *really the earliest* forms of life on our globe, necessarily destroyed and burnt up as they must have been in the metamorphic and igneous rocks. And this consideration is, again, closely connected with those profound suggestions thrown out by one of our most eminent philosophers in relation to the deposition of strata in connection with the effects of central heat; who observes — "According to this view of the matter there is nothing casual in the formation

* As connected with these points, and more especially the development of fishes, the reader will find some able illustrations, refuting many of the minute details of the theory of the "Vestiges," in a very able article in the British and Foreign Medico-Chirurgical Review, No. 26., April, 1854, p. 425.

But however open to criticism may be the details of particular physiological statements in the work referred to, the whole tenor of the preceding discussion will show the degree, in which I cannot but concur, in its broad philosophical principles.

of metamorphic rocks. All strata once buried deep enough (and due TIME allowed!!!) must assume that state,—none can escape. All records of former worlds must ultimately perish." *

Regular laws from unity of type.

But the invariable relation of all the successive forms to one primitive type, constitutes the legitimate and undeniable evidence of some regular order of causes presiding over their production, operating through periods of time of enormous length, during which old species have slowly disappeared by the action of natural causes, and new allied species have as gradually appeared beyond all doubt as much in accordance with other equally natural, even if at present unknown, laws — parts of the great order of causes, in conformity with which these and all possible physical events must have taken place.

Recapitulation.

In what has preceded it has been, I trust, sufficiently shown that some of the arguments most commonly adduced, whether in support of the "immutability of species," as supposed analogous to the permanent laws of nature, or against their "mutability" as "contrary to experience," or in favour of

* Letter from Sir J. Herschel, in Mr. Babbage's 9th Bridgewater Treatise, Appendix, p. 240. 2nd edit.

interruptions of natural order from apparent gaps in the geological series, are all destitute of foundation, fallacious, and untenable.

On the other hand, while these arguments, which are those most commonly relied on against transmutation, are in my opinion completely refuted, there is still no *positive evidence* to establish it as *a demonstrated theory*. Yet as a mere *philosophical conjecture*, the idea of *transmutation of species under adequate changes of condition*, and in incalculably *long periods of time*, seems supported by fair analogy and probability.

No necessity to suppose interruptions of order.

Taken for what it is worth as a conjectural hypothesis, it may be regarded as helping the general conception of *some great principle of orderly evolution*, according to which the present as well as past systems of existence have been produced out of preceding orders of things: and as at least conspiring with all truly philosophical considerations to disprove the necessity for appealing to any sudden interruptions of order, or operations of an unknown and mysterious kind, alien from all natural causes.

Conjectural hypotheses in subordi-

It should, moreover, be carefully observed that, though particular hypotheses of this kind may fairly be indulged in, regarded as such, yet an exclusive devotion to them may be prejudicial to the firm grasp which the mind should rather seek to maintain of the *broad principle*, the subordination of all events to *some general laws*, however at present *undiscovered*, based on the maxim that throughout nature *what we do not know must really be as much under the dominion of law as what we do know.*

nation to broad principles.

True science is always ready to confess the failure of *existing* means of investigation when a limit appears placed upon its advances, coupled, however, with an assurance that that limit will some day be passed.

It is legitimately within the province of inductive philosophy to suggest conjectures as to the operation of grander laws of vitality acting through the immense periods of past duration, and of which, during the brief and momentary duration of existing things, we enjoy only the most partial, imperfect, and occasional glimpses.

Permanence of unity of composition.

It is eminently consistent with the great principle of the uniformity of nature through all time, to suppose that like many lesser laws in the natural world, that of the *existing* permanency of species, may yet be subordinate to a greater and more comprehensive law of change, requiring such vast periods for its accomplishment that no measurable portion of time may suffice for the production of a sensible amount of variation.

Having thus far considered the general principles, I will proceed briefly to glance at the more particular views of some writers.

Narrow views prevalent.

Looking at the question in a perfectly dispassionate manner, there appears to me a one-sidedness in the censures, or at least, excessive cautions, often expressed, against so hazardous an hypothesis as that of transmutation, even by some eminent philosophers, more than is warranted by sober philosophical considerations; and in which others display more zeal than can be explained by mere antagonism in a fair scientific controversy, while they sometimes appear to betray even a degree of alarm at the bare suspicion of a leaning towards the obnoxious theory of development, as if their whole scientific, or even personal reputation were at stake.

Polemical spirit of some discussions.

Some, again, have taken up such questions in a more determined controversial spirit, and have maintained in a tone of polemical acrimony, little to have been expected on such a subject, that the phenomena of new species are absolutely impossible to be explained on any physical principles, or even by any physical conjectures; and must be ascribed to sudden interruptions of the order of nature, connected with the convulsions and catastrophes which overwhelmed all the old species, and were of a kind wholly beyond the domain of physical causes or the limits of philosophical examination.

Dogmatic assertions.

Such imaginations easily find favour with those who have *some other object in view than mere philosophical truth;* and if somewhat faulty in their foundation, their weakness in reason is abundantly compensated by loudness of dogmatism and a peremptory style of assertion that "species are real existences," and that "transmutation is impossible;" all which has an imposing effect when supported by the aid of a kind of mystified eloquence, and seconding the more awful denunciations so authoritatively pronounced against the heterodox speculations of the developmental school.

Theories of sudden origination.

But when (as we have observed) some of the opponents of transmutation do not content themselves with mere negation, but assert another theory of sudden originations of animal life, we have clearly a right to demand of them some distinct statement *of their own meaning,* some definition of the nature of the theory they propose to substitute, and the *process* by which *they* conceive the results to have been brought about.

Let us, then, imagine the case in question; among other allied species already existing let us suppose that an entirely different *and truly new species* has *suddenly* made its appearance. The individuals of this new species are found living and growing by ordinary means, and in all respects subject to the same regular condition as everything around them. It is the fair right and object of the inductionist to ask *how long* have they continued in this state? Did the natural process of growth reach back to their evolution from a seed or an ovum? or to *what stage* of early existence or rudimentary evolution?

Altogether fanciful and unintelligible.

If not derived from a parent, was the ovum formed out of its component elements already existing in matter around? Was the organism gradually evolved, or is it meant that it assumed its form suddenly, or that the entire creature started into existence, full grown, out of the earth, as in the frescoes of the

Vatican or the imagery of Milton*; or that it assumed a palpable existence out of nonentity?

If such be their meaning, all that the inductive philosopher asks is to see some slight *proof* for these marvellous assertions, *which he will be quite prepared to admit, if sufficiently verified;* but must, after all, ascribe to some action of regular physical causes as yet unknown. The "onus probandi" clearly lies on those who assert such extraordinary hypothesis, and not on those who, if they indulge in any speculations of the kind, are careful to found them strictly on probable analogies, dependent on the great laws of unity of composition and modification of parts; but in every case strictly conformed to the one grand overruling principle, the universality of law, order, and continuity, presiding as powerfully over the earliest stages of creation as during its continuance at the present moment, and applying equally to organic as to inorganic existence.

Necessity for proof.

Thus some scientific inquirers reject the idea of development or transmutation, because, as they allege, they find no evidence or existing instances of

Inconsistency of such views.

* Par. Lost, vii. 463.

such a thing to produce; yet, in its place they assume a sudden production of full-formed animals out of their elements, or out of nothing, which is still farther remote from any possibility of proof from experience, and even beyond all rational conception. They discard one theory for want of proof, though easily imagined and understood, only to adopt another which equally wants proof, and is at the same time wholly unnatural and incapable of comprehension.

Fallacious illustrations.

It is but putting the fallacies above refuted into a variety of vividly illustrated forms which constitutes the staple of some very popular writers. We find it, for example, pervading the work of Mr. H. Miller * (already referred to). To take a single instance, we may cite that striking passage where he so graphically describes the lake of Stennis, partly salt, partly fresh, partly brackish, and observes that each portion has its appropriate species of plants and animals, which *never have been found* to exhibit any signs of change, migration, or intermediate mutations, *so long* as they have *been observed*, and the *conditions* of its several

* Footprints, &c., p. 240.

parts have remained the same; whence we are to conclude that no such mutations ever are, have been, or can be made, under any conceivable changes of condition, and during any period of time, however incalculably great!

The same writer upholds the truth of the gradual "elevation of the types of being in the successive stages or conditions of the earth, as corresponding to those changes in the dwelling-place assigned to the animal creation, and suited to their successively improved natures."* And it is his main object to maintain that these changes were all of an isolated nature, and that they all happened in an abrupt and unconnected manner, inexplicable by natural causes; and this (it would seem) solely on the ground of the alleged breaks or interruptions (as he considers them) between different formations, which were before considered.†

As to the mode in which these changes were brought about, the author appears to proceed on the ground that *all* scientific modes of explanation *must fail*, because *one* such mode (in his opinion) fails; he

* Footprints, &c., pp. 283. 286.

† See above, p. 340.

therefore takes refuge in the assertion of immediate and repeated sudden "*creations*," as the only way in which the continued production of new species ought to be spoken of, and even strongly denounces all attempts otherwise to explain them. And (unless I misapprehend the author's meaning) it would seem as if he seriously upheld the notion of animals being thus produced *full grown*, since he expressly considers Oken's theory of development out of a monad or infusorial point, sufficiently refuted by referring to the existing *size* of the gigantic Asterolepis*, and other fossil remains, which he assumes to have been thus suddenly produced; as if, because they had *grown* to that size, these creatures had never been in an embryonic state, perhaps microscopically minute.

Physical order in all past changes.

Legitimate science can never lead us to anything but higher generalisations of physical order; it can never point to operations of a kind beyond regular causes, or warrant a reference to hypotheses stamped with the professed character of mystery and inscrutability, at an earlier, any more than at a later epoch; in the primeval arrangements any more than in the

* Footprints, &c., p. 119.

existing maintenance of the organised world. Where we confess ignorance of intelligible truth, it is at once absurd and presumptuous to create inscrutable mysteries. It is clearly preposterous to maintain that an ancient event ill understood is an absolute deviation from all natural order, merely because we cannot at once interpret it; or that it is beyond all physical causes, because we do not daily see instances of its occurrence.

Explanation of changes no explanation of first origin.

But, lastly, if it were granted that we could follow up the successive development of species even to a *very* remote date, it must be evident that this would still carry us but a little way towards *the real first origin of all things*, and would manifestly be but a single step in the course of tracing backwards the order of creation. The question even of the *first origin of organised life* would still remain. In reference to *successive forms and changes* of life we have the evidence of existing remains as the basis of our reasonings. But we have no such evidence of the *beginning* of the series. Even if the idea of transmutation or any equivalent principle were granted as to comparatively later variations, still the question would arise, What and how many were the original types from which these unlimited varieties began to diverge? and how did the primitive germs of life themselves originate? On these points theories have not been wanting.

Theory of Lamarck.

Lamarck represents nature as continually engaged in the gradual formation of the elementary rudiments of all animal and vegetable existences of the simplest kinds, which are afterwards compounded into more complex forms. These rudiments or monads are the only things to which she gives birth directly. He regards them as probably of a distinct kind for each of the great divisions of the animal and vegetable kingdoms, and supposes that they are gradually developed, but subject to material modifications from the action of external causes.

Theory of Oken.

This theory is closely allied to that of Oken*,

* That remarkable work, the "Elements of Physio-Philosophy" of Oken, has attracted some notice in this country, through the translation of it published under the auspices of the Ray Society (1847). Its nature is confessedly speculative and hypothetical. The author expressly tells us in the preface that he wrote it off under "a kind of inspiration," and more than once refers to its deficiencies in proof from matter of fact.

With regard to the primary principle of organised life the author declares it the object of his speculations to show how, by self-evolution of the elements into higher and manifold forms, they become finally organic, and in man attain to self-consciousness. Man includes the representation of all lower forms, and these again are but man disintegrated. (§ 10—19.) The grand principle is the origin of all organised life from an infusorial cell, formed out of what the author terms the "primary mucus," "schleim-substanz," or protoplastic matter, which, he says, is "carbon mixed identically with water and air." (§ 898.) Decomposition is only a transition from one life to another, which takes place through this mucus, into which organised matter is redissolved. "Every which, avowedly of a speculative kind, and bearing a metaphysical aspect in its first principles, is yet put forth as a physical generalisation; though from the very abstract nature of the ideas and language employed it is difficult to estimate the precise evidence on which it is supported. The author's main principle, however, appears to be the origin of all organised life from a primary infusorial cell or monad, formed out of an elemental substance, which seems to be simply a compound of the admitted inorganic elements of animal matter. From the aggregation of such cells the various organic structures are compounded, and this he seems to regard as a process constantly going on in nature.

It is sufficient here to glance at such speculations, all essentially hypothetical; these and others of the like class are at least of use in serving to show that there is nothing *impossible* or *inconceivable* in the idea that such production of the rudiments of new life

generation is a new creation." (§ 924.) Of this mucus a cell, cyst, or vesicle, is formed, called "infusorium," the primary germinal principle in plants and animals. (§ 930—943.) The author says, " No organism has been created of larger size than the infusorial point, which is microscopic: all larger forms are developed, not created. By the aggregation of such vesicles organised forms arise, which by successive combinations of the more simple, produce ultimately the higher and more complex structures. (§§ 958. 2961. 3161. 3175.)

may have gone on in accordance with a regular physical system.

Inductive inquiry does not extend to beginning of all physical causes.

However high we may ascend in the order of time, it is still perfectly within the province of physical inquiry to endeavour to unfold the steps in the process by which changes in the order of things have been brought about. It is within the limits of philosophical conjecture to speculate on the gradual accomplishment of the great design of educing the existing order of the universe out of former conditions, in whatever degree of obscurity the precise *modes of action* may be enveloped by which it was effected.

As far as we can trace backwards, even in imagination, the succession of events into the depths of primeval time, we can only conceive them succeeding one another in determinate order, by the operation of profoundly adjusted causes, whose nature becomes less and less imaginable to us as we recede into higher antiquity; but we cannot draw any line, and say here the series began, or here all continuity commenced.

All ideas of a beginning from other sources.

To attempt to go back in imagination to an epoch prior to all the changes of matter at which it was first constituted or called into being, is what no inductive philosophy can warrant. No analogy points to such a beginning of physical causes. Physical philosophy always supposes at least some physical elements in existence; it cannot investigate or conceive a condition antecedent to nature, or the case of its actual commencement. No science can carry us, even in imagination, into a state of arbitrary and disordered influences; a *chaos* has no existence in the ideas or the vocabulary of the inductive philosophy. A creation, in the same vocabulary, implies orderly evolution. If we entertain any ideas beyond these, it can only be from sources of quite another kind.

§ IV.—THE BEARING OF THE PRECEDING ARGUMENTS ON THE THEOLOGICAL VIEW OF CREATION.

Theological view of the inquiry kept distinct.

It was observed at the outset of this Essay, that the question of Creation has distinctly a theological bearing; and it is no disparagement to such a view, that in the preceding sections I have treated the subject in a purely inductive and scientific light, and have *purposely abstained* from introducing any reference to those higher considerations, in order to lay a more secure basis for any such applications, as well as for meeting any objections alleged on religious grounds.

Prejudice against such speculations on religious grounds.

It cannot be denied that any discussion of the question of Creation, or any attempt to trace the probable history of the origin of the physical world, or of its organised productions, on merely scientific grounds, has been often regarded, especially by a certain class of minds, as having a tendency unfavourable to religion, and as being, in some degree, an intrusion into its province and an assumption of its office. Such impressions, however, appear to me to take their rise in the same common species of misconception of the relations in general between science and faith, which, in so many other instances, has resulted either in a lamentable antagonism and hostility, or in futile attempts to combine them in incongruous union, upon fallacious principles.

Arising from misconception.

I have, in another place*, considered some instances in which the discoveries of science are undeniably at variance with doctrines which had become identified with popular belief, or had even been erroneously received as part of the established creed. And when any topics, having a similar bearing, come into discussion, if, on the one side, they are naturally taken up with the zeal of religious prepossession, on the other, they are often not fully examined; they are impatiently dismissed, or thought sufficiently treated, if glossed over by a few vague, specious, and evasive generalities.

There exists, unhappily, too great an unwillingness on either side to meet such questions with perfect

Want of open discussion.

* See Essay II. § ii.

honesty and fairness. The astronomer, the physiologist, or the geologist, for example, may be fully enlightened as to the extent to which some of the conclusions of his own science may clash with certain received articles of popular belief. But, devoted to that science, and caring more to relieve it and even himself personally, from hostile insinuations, than to promote any higher views of truth, he more naturally than philosophically seeks to conciliate the matter in an ambiguous phraseology; as if accepting literally the irony of Lucian, who, after relating a story of a philosopher having been maltreated by a mob for attacking some of their superstitions, adds,—"And very justly; for what right had he to be rational among so many madmen?"

Suppression of discussion injurious to religion.

But still more injurious to the cause of religious truth is the course too often resorted to by the professed defenders of its cause, even in the present time. Not always duly alive to the actual spread of intelligence, they cringe to the loud but ignorant zeal of the few, and become followers in the train of prejudice rather than its correctors and enlighteners. They have too often yet to learn that, by continuing to insist on dogmas which the advance of knowledge has discredited, and literal interpretations which the discoveries of science have set aside, by adopting fallacious compromises, or by discouraging and denouncing those open avowals which alone consist with the reality of truth, and that free inquiry which Christianity challenges—they are following a course as unworthy in principle as it is short-sighted in policy; they are inflicting the worst injury on their own cause, and are but strengthening the arms of that sceptical hostility which they so strenuously profess to oppose.

On the other hand, here, as in other subjects, resolutely yet cautiously, to pursue the free course of rational inquiry—and therein to follow *truth*,—we may be assured can never lead to *evil;* while every advance in real enlightenment, in proportion *as it is real*, must of necessity cast its beneficial rays equally over science and over faith.

Evidence of science to the supreme mind, not to the act of creation.

In the preceding discussion, we have seen that physical inquiry traces the development of the existing material world up to a certain point, and allows us to conjecture a few stages beyond that; but where analogy ceases to apply, or conjecture to find materials, there all *physical speculation simply termi-*

nates: it does *not* even point, however obscurely, to any event beyond. Its province is *Nature* in space and in time. With any class of ideas *of a different kind* not referring to *Nature,* it has no concern. The inference, indeed, from all physical truth which impresses us with the idea of *Omnipresent Mind,* is one everywhere presenting itself, as much in every phenomenon at the present moment, and in our actual locality, as in the remotest points of distance in time or space. The physical evidence of *Creation* is universal, but it does not indicate any one point as its commencement.

The proofs to which I have referred in a previous Essay* of the great Moral Cause of the universe, are of a kind related rather to *permanent* and enduring evidence of the order of things, than to any inferences as to special *past acts* or *events,* which may be imagined to have been its direct manifestations at remote epochs: but any such supposed events would afford similar evidence only in proportion as they might be found evincing a conformity to some high principle of order and law. The evidence of the material world points to

* Essay I. § v.

ever-existing Mind continually manifested in the existing order of nature, as well as in all the successive changes which we can trace through countless periods of past time; but which, in all their varied modifications, present not the smallest deviation from one great type of unity and harmony.

Belief in a beginning from other authority.

In a word, the idea of *a beginning of Nature in time* is one which no *physical* philosophy can teach us. It is an idea wholly derived from other considerations. There are metaphysical arguments as to the impossibility of conceiving eternal matter, because (it is alleged) it must then be self-existent, and the like. But, without entering upon them, it will suffice to observe, that all such reasonings are of very different degrees of force to different minds, and perhaps have little effect on the generality. In point of fact, by far the great majority have obtained their idea of "Creation," not from any such arguments, but from the prepossessions of early instruction, by which that term, with a certain religious meaning affixed to it, has from childhood been incessantly impressed on their ears and memory,—though often but little deeper.

As to any speculations of science on the earliest

Nebular theory accordant with the idea of design.

history of the formation of the world, such as the nebular hypothesis, or any like supposition, in reference to our present view of the subject, there is only this to be said: in proportion as any such theory (supposing for the moment it were established) suggests an *orderly evolution* according to *preordained* physical laws, *so far* it keeps in harmony with the idea of *Supreme Intelligence*, and thus implies a worthy notion (as far as a mere conjecture may avail) of creative power and wisdom. If we could in imagination trace the supposed process through any antecedent stages in a similar way, it would in proportion afford a similar extension of such reflections.

It suffices to allude to such hypotheses in connexion with the present subject, merely for the purpose of remarking, on the grounds now referred to, how entirely futile and irrational are those charges often brought against theories of this kind, that they have an irreligious tendency.

Subsequent general arrangements of our globe indicative of design.

When from such remote contemplations we descend to the comparatively more accessible inquiry into the *changes* which have taken place in our globe and its organised inhabitants, it is fully admitted that the farther these successive changes have been investigated, and traced to determinate laws,—the more links in the chain we can unravel,—the more they are always found to disclose the evidences of creative wisdom and design. But, when we speak more precisely of "Creation," in the sense of the commencement of organic life—and especially of the introduction of new forms of life,—it becomes more necessary to examine the bearing of the theories of their origin on the cause and argument of religion, whether natural or revealed.

Views of origin of life and of new species in relation to the same argument.

No real interruptions.

Since on the high principles just referred to, the truly inductive philosopher cannot suppose any such idea as that of a real deviation from unity of plan, so he cannot but feel assured that such an event as the introduction of a new species, or even the first origination of life, could be nothing of an arbitrary kind, but must have been part of the great order of preordained causes whose pervading influence essentially distinguishes CREATION from CHAOS; and in which any apparent interruptions can only arise from our confined apprehensions of the vast scheme, essentially *one*.

The evidence of palæontology throughout such

Unity of plan through incalculable periods of past time.

inconceivably vast periods, is of the most overpowering force. The simplest contemplation of the facts of ancient organic life shows that even those forms of the earlier epochs which are the most dissimilar to any existing, are all of *one family* with them. Throughout these unfathomable depths of primeval time, which it transcends imagination to conceive or arithmetic to express, the organic world is, and always has been, emphatically *one:* modelled on one plan, and amid all diversity exhibiting one common feature of a grand *recondite and comprehensive unity of design*; or as Œrsted has expressed it, "The animals and plants of former periods are all different emanations from the same great thought."

Equally emphatic are the opinions expressed on this point by some of our most eminent men of science.

In continuation of a forcible passage before quoted, Professor Owen goes on to observe, "To what natural laws and secondary causes the orderly succession and progression of such organic phenomena may have been committed, we as yet are ignorant. But if without derogation of the Divine Power we may conceive the existence of such ministers, and personify them by the term *nature,* we learn from the past history of our globe that she has advanced with slow and stately steps, guided by the archetypal light, amidst the wreck of worlds, from the first embodiment of the vertebrate idea under its old Ichthyic vestment, until it became arrayed in the glorious garb of the human form."*

To this noble passage I cannot forbear adding the single comment that, according to my view, not only "without derogation of the Divine Power," may we entertain the ideas so beautifully expressed; but, if there be any truth in what has been before advanced, so far from anything *derogatory,* such a view constitutes the *very proof* and manifestation of that power, and is just what enables us legitimately to trace its operations — as alone we can worthily trace them — in the indications of law and unity, order and system; while without such evidences of Universal Mind and Supreme Reason, arbitrary intervention might be only irresistible fate, and sudden revolutionary changes and convulsions only atheistic anarchy.

* On Limbs, p. 86.

Again, it would, perhaps, be impossible to find a more truly admirable exposition of the case than in the following sentence from the pen of one of our first living philosophers: — "For my own part I think it an inadequate conception of the Creator to assume it as granted that His combinations are exhausted upon any one of the theatres of their former exercise; though in this, as in all His other works, we are led, by all analogy, to suppose that He operates through a series of *intermediate causes;* and that in consequence, *the origination of fresh species,* could it ever come under our cognizance, would be found to be a *natural,* in contradistinction to a *miraculous,* process; although we perceive no indications of any process actually in progress which is likely to issue in such a result.*

Evidence of creation in physical evolution.

As in the natural world the only indications we have of the operations of the Divine mind are those manifestations of order; so whatever we ascribe to the same source we can only conceive as worked out in accordance with the same great principles.

The real question in any such cases is not whether

* Letter from Sir J. Herschel, in Babbage's 9th Bridgewater Treatise, Appendix, p. 226. 2nd edit.

certain events or processes are or are not to be traced to the Divine will and counsels — for that is not denied by any reflecting inquirer — but simply whether the *mode and method* of the Divine operation can be either absolutely discovered, or even reasonably conjectured, to have proceeded in this or that particular path.

Question not as to *fact*, but the *mode* of Divine operation.

A rational physico-theology teaches that the succession of forms of organised life on the globe, up to the first origination of all animated nature, were acts of the Divine will wisdom and power, in precisely the same sense as the revolutions of the double stars and planets, the daily tide, the fall of rain, the ascent of vapour, the action of the sun's light and heat, and all other natural phenomena, regulated by similar recondite laws, are direct and immediate acts of the same Divine will wisdom and power.

Creative power constantly manifested.

And, indeed, to approach still nearer to the idea of *origination* or *production*, we may find creative power as strictly and properly exemplified every day in the marvellous process of evolution of animals and plants out of a mere microscopic germ or embryo, as in any events of past times. Those events may perhaps appear preternaturally magnified to our intellectual vision from the medium of unfathomable antiquity through which we view them; but we may be assured that simplicity is as sure a mark of the Divine operations as grandeur; and that equally in the present as in the remotest epochs of the past.

But there are some particular instances which have been often dwelt upon as putting the question to a more distinct test, and which may be perhaps more peculiarly suggestive to those who have not been prepared to take the more enlarged view of the subject on first principles.

Instance.

To take only a single instance, rather by way of illustration, we may refer to the multitudes of infusoriæ and animalculæ whose existence is *restricted and peculiar to artificial products* made by man, and must have been consequently introduced at dates not only *subsequent* to that of man, but continually recurring.

Now, the question of their origin involves equally remarkable consequences whichever way it is viewed. If they are "developments" of existing allied species, only modified so as to suit the particular conditions under which they exist, a principle is conceded which cannot be consistently refused in other cases. If they are the result of special interventions to bring them into existence out of nothing, they constitute such a multiplication of miracles as the most strenuous advocate must disavow; and after all according to all acknowledged principles, a miracle so *continually* and *regularly* repeated ceases to be a miracle.

Objections to physical theories of evolution.

Theories of the physical evolution, or origination, of new species in past epochs of the earth's history have been specially the objects of censure and denunciation to some eminently religious writers; and they have been sometimes condemned with a degree of warmth and violence which clearly indicates the admixture of a larger element of religious prepossession than of reason.* But even in a religious

* Such accusations are often very unfounded. But when they are imagined subservient to a sacred cause, truth and fact, justice and candour, are too commonly looked upon as secondary considerations, or rather the disregard of them would seem to be considered as only the more praiseworthy evidence of religious zeal.

One of the most striking instances of this kind of religious vituperation at the present day has been that of the incessant attacks made on the "Vestiges of Creation," on the alleged ground of the impious and atheistic tendency of its speculations.

Now whatever may be thought of the theory or speculations, as such, nothing can be more utterly and palpably unjustifiable than the charge of an *irreligious* tendency against a work *in which almost every page is replete with expressions of the most devout homage to the Divine power, wisdom, and goodness.*

point of view it is, in truth, by no means easy to see on what substantial grounds such vehement opposition can have arisen.

Objections unreasonable.

So far as the evidences of *natural theology* are concerned, it would follow, from all that has been here advanced, that those evidences, so far from being discredited, could but receive increasing confirmation in proportion as *any physical theory* might be substantiated, which would give us a deeper insight into those secondary processes and laws by which the Divine will and plans were worked out, and which are to us the manifestations of such designs.

Confusion of ideas.

But, perhaps (as in other instances elsewhere noticed*), nowhere has the confusion of thought respecting "causation" been more misleading than in reference to the subject of *Creation*, giving rise to the constantly reiterated but absurd accusation against all theories which aim at tracing the series of physical events as far back as possible towards the origin of things, that they fly to "second causes" in the desire to avoid the acknowledgment of a

* See Essay I. § v.

"First Cause," or that they endeavour to get rid of a Creator because they seek to trace more in detail the steps by which His work is carried on.

Such often refuted cavils are, however, constantly revived by a certain class of minds; indeed, every advance in discovery or philosophic speculation from the days of Galileo downwards, has been, as a matter of course, accused of having an irreligious tendency.* Every successive step made good in bringing Nature under the dominion of law, is stigmatised as setting up Nature instead of God,—as if we could trace Him except through Nature; and as referring everything to mechanical causes instead of the Divine will,—as if the recondite system of physical causes were not the very evidence of that Will and that Supreme Wisdom.

Bigotry.

And we may observe, that if the supposition of original adjustment superseding continual interposition be objectionable in the instance of a succession

The same objections must apply to all physical theories.

* With a certain class of religionists every invention and discovery is considered impious and unscriptural—as long as it is new. Not only the discoveries of astronomy and geology, but steam, gas, electricity, political economy, have all in their turn been denounced; and not least, chloroform. Its use in parturition has been anathematised as an infraction of the penalty pronounced on Eve!

of varied forms of species, evolved by some processes and in accordance with some fixed law as yet unknown, it must be equally so when evinced in other instances better understood; for example, in the "stability" of the planetary system, the "conservation of areas," or even the very "inertia" which keeps up the planetary revolutions, or, indeed, throughout the whole system of the physical world referred to an invariable system of laws originally imposed, and by the combinations of which all actual phenomena are brought about. If perpetual intervention and constant new volition be the only religious view of the matter, we ought to discard Laplace and Newton and go back to Kepler's "vital forces" and spiritual beings spinning the planets in their orbits, or Descartes' World of Vortices animated by the Divine Soul.

Objection as leading to Pantheism.

Others have denounced all theories of physical evolution as leading to and implying *Pantheism*, an accusation, if possible, more strange and groundless. The theological question between development and successive "creations" is simply whether we suppose the Creator to construct a machine which, once adjusted, shall go on fulfilling its work, or one which at successive periods shall require repeated manual interposition. The Pantheistic theory (to carry on the same metaphor) would assert that the machine was self-animated, or was itself the artificer or regulator of its own parts and movements. The first idea (at least rational and worthy) has manifestly nothing in common with the latter absurdity.

Objections to development theory on doctrinal and scriptural grounds.

But the main source of the difficulty and objections which have been felt, on religious grounds, against any theory of the evolution of organised existence by the agency of natural causes, is their supposed repugnance to particular views of theological *doctrines*, and the declarations of *Scripture* on which those doctrines are founded.

Discrepancy between Scripture and geology.

In a former Essay* I have adverted to the question of discrepancies between science and the language of Scripture generally, and have referred more especially to that notable instance of it—the irreconcileable contradiction between the whole view opened to us by geology, and the narrative of the Creation in the Hebrew Scriptures, whether as briefly delivered from Sinai, or as expanded in Genesis. In the minds

* Essay II. § ii.

Even where admitted, other doubts remain

of *all competently informed persons* at the present day, after a long struggle for existence, the literal belief in the Judaical cosmogony, it may now be said, has died a natural death. Yet many are still haunted by its *phantom,* which perpetually disturbs their minds with apprehensions equally groundless on collateral points.

Most rational persons now acknowledge the failure of the various attempts to reconcile the difficulty by verbal interpretation; they have learnt to see that the "six days of thousands of years" have, after all, no more correspondence with any thing in geology than with any sane interpretation of the text. And that the "immense period at the beginning," followed by a recent literal great catastrophe and final reconstruction in a week, is, if possible, more strangely at variance with science, Scripture, and common sense. Yet, while they thus view the labours of the Bible geologists as fruitless attempts, they often do not see that they are fruitless, not because they fail in detail, but because they proceed altogether on wrong grounds and in a wrong direction, and thus remain under the dominion of the same radically mistaken prepossessions, which lead to not less unhappy misconceptions on other allied topics.

Well might Humboldt* speak of geology as "now finally abstracted, *on the Continent, at least,* from Semitic influences." But in this country it may be hoped a better epoch is beginning to dawn, as it must do, in proportion as men reflect on the real basis of their reasonings, and learn to apprehend clearly the distinct grounds on which science and Christian belief respectively repose.

Evolution not more opposed to Scripture than all geology.

And if to the general truth of the immense continuous series of slow and gradual formations constituting the earth's crust, disclosed by geology, we add the grander theoretical inference, that all the varied modifications of animal life were equally produced according to some regular scheme of physical causes; or, if the more imaginative speculator should think that he can identify that scheme with certain physiological indications of rudimentary evolution,—it is impossible to see in what respect the latter class of views can affect religious considerations *more* than the former, or be *more opposed* to the letter of the Mosaic

* Cosmos, 1st transl. p. 288.

description than they are. Yet there are those who seem to view these last ideas with more peculiar apprehension. The discrepancy cannot really be greater whether we adopt any physical theory of the mode of origination of successive forms, or whether we reject all such speculation. In a word, those who accept geological truths at all, and admit the palpable contradiction to the Old Testament, without prejudice to their faith, cannot with *consistency* make it a ground of objection to any hypotheses of the *nature* of the changes indicated, that they are *contrary to Scripture*. They are in no way *more* so, than *all geology* is.

Idea of creation out of nothing.

The idea which is often attached to the word "Creation," as meaning a calling into existence *out of nothing* (as already observed), rests wholly upon certain *metaphysical* arguments* which it is no part

* The doctrine of the origination of matter *out of nothing* has been upheld on various metaphysical grounds; but it has received peculiar support from the speculations of Oken, in his work before referred to. (Elements of Physio-Philosophy, by Lorenz Oken, M.D., Prof. Nat. Hist. Zurich. Transl. by G. A. Tulk, M.R.C.S. Published by the Ray Society, 1847.

His whole theory rests, in the first instance, on what is termed "Mathesis" and "Ontology," which consists in deriving the original existence of matter on the principle of the algebraic formula $0=+-$, or "out of nothing there is something," which is applied in an hypothesis of antagonistic principles of existence, under certain conditions neutralising each other, but under others evincing independent existence.

From metaphysical principles of this kind the author deduces the of my design to discuss. But with reference to the opinions of those who lay so much stress on the *letter* of the Bible in such points I would merely observe that, whatever may be the value of other arguments in support of it, it is wholly destitute of any foundation in *Scriptural* authority. The word which in Genesis and elsewhere, is rendered "create," has been pronounced by eminent Hebrew scholars* by no means to bear the sense above mentioned, being only a stronger or *more intensitive* form of expression of the idea of *making* or *fashioning*. While other passages leave the idea at least equally indefinite; if, indeed, they do not in some sense refer to pre-existent matter.†

---

existence of an "eternal self-consciousness, which is God" (§ 61.), and even the *mode* of his existence leading to the doctrine of the Trinity (§ 67.).

Again, on similar grounds, introducing the agencies of heat, light, ether, and especially "polarity," from a combination of these with the antagonistic ideas before mentioned, he derives monads, and thus the world's development out of nothing by processes which, he says, are the exact "Genesis of Moses," terminating at length in the creation of man. (§ 958.).

He ascribes the existence of the universe in the first instance to the Divine will: "God spake, and it was." (§ 63.) And, again, "God has made heaven and earth out of nothing. God has not found matter co-eternal with Himself, and, like an architect, arranged this to His fancy; but he has out of His own eternal omnipotence by His will simply evolved the world out of nothing into existence." (§ 167.) But all this will not satisfy the bigots, who set down Oken as an atheist!

* On this point the reader is referred to the authority of Dr. Pusey in Buckland's Bridgewater Treatise, i. 24.

† See *e.g.* Heb. xi. 3.

Organised beings not formed out of nothing.

Moreover, in the particular instances of the creation of animals and plants, it is notorious that in the very language of the Mosaic narrative, the aquatic animals are described as being brought forth *out of the waters**, as the land animals and plants are *out of the earth,* and man especially, as formed *not* out of nothing, but out of "the dust of the ground." Thus, at any rate, those who maintain that the first individuals of all new species were always brought into existence out of nothing, can at least have no shadow of *Scripture* authority for such a belief.

Inconsistency of repeated creations.

But the most strange and inconsistent part of the whole seems to be, that those professing such literal adherence to the Mosaic narrative should utterly disregard it in having recourse to so many successive *repetitions of the work of creation* in different epochs, when the whole drift and purport of that narrative is manifestly and palpably directed to the one special object of representing the whole as a single creative act, *begun* and *completed* in the six natural days, with peculiar and emphatic reference to the *final cessation* and rest on the seventh. Equally inconsistent are

* Gen. i. 11. 20.; ii. 7.

those who contend for the primeval Sabbath, and yet uphold the seven periods of unlimited length!

Scriptural representations adapted to the state of knowledge.

The prevalent theology is too deeply immersed in an indiscriminate and unthinking Bibliolatry. But even on the fullest admission of inspiration, the slightest rational reflection must show the unreasonableness of looking for indications of the inspired character of Scripture, in relation to any other subjects than those of its proper spiritual communications; and even these in the *mode* of their introduction are always specially *adapted* to the apprehensions and condition of those to whom they were addressed, and always to be applied subject to the due discrimination of circumstances, times, parties, and dispensations.

Reference to creation in the New Testament.

Thus, more precisely with respect to the subject of "Creation," the writers of the New Testament, doubtless adopting themselves the existing belief respecting it, yet never dwell upon that belief in detail*, nor insist on any of its peculiarities. They refer to it, in fact, only in a general sense as opposing

* We may except one solitary instance (an exception which eminently proves the rule), when the Apostle (Heb. iv. 4.) is specially arguing with *the Hebrews*, and refers to their belief in the Divine rest on the seventh day as figuratively applied to the future and everlasting rest of the faithful.

the superstitions of heathenism*, and teaching the Gentiles that the elements of the material world, which, either directly or under various mythical personifications, had been the objects of their worship, were, in reality, the *creatures*, not the *Creator*, to whom alone worship was to be given. The only specific references made, are those of a more elevated and mysterious nature, involving *no physical ideas*, but referring the work of creation to the Divine Logos†; probably in refutation of the speculations of the Gnostics; or, as some think, in giving a Christian sense to them.

Belief in the recent date of man's origin.

To the same kind of misapprehension may be traced—but even with less appearance of reason—the zeal with which the belief in man's *recent origin* on the earth has been maintained, and the suspicion and animosity excited by even a hint or conjecture at any possible higher antiquity of the race. The prevalent belief in the very recent origin of man, geologically speaking, depends wholly on negative evidence. And there seems no reason, from any good analogy, why human remains might not be

* Acts xvii. 25.
† John i. 1.; Col. i. 16. See Dr. Burton's Bampton Lectures, p. 112.

found in deposits corresponding to periods immensely more remote than commonly supposed, when the earth was in all respects equally well suited for human habitation. And if such remains were to occur, it is equally accordant with all analogy to expect that they might be those of an *extinct* and *lower species.* The only real distinction in the history of creation which marks a supposed "human epoch" is not the first introduction of the *animal man* in however high a state of organisation, but the endowment of that animal with the gift of a moral and spiritual nature. It is a perfectly conceivable idea that a lower species of the human race might have existed destitute of this endowment.

From the Hebrew chronology.

The belief in the recent date of man is usually adopted from the received Hebrew chronology, itself (as is well known) open to critical difficulties. But, indeed, to those who imagine the Old Testament authoritative in matters of philosophy or chronology, there is no limit to inferences of this kind. There are some, even, who believe that the "permanence of species" is a *Scriptural* doctrine, because it is said that plants "after their kind" "have their seed in themselves!"

Objection as to the origin of man.

But the idea of a physical process of origination of organic life has excited a more peculiar opposition, on the ground that it would include MAN *and his descent* in the general category, and represent the human race as at some remote period gradually developed out of an inferior species, which, it is alleged, savours of materialism, and lowers the moral dignity of man. Now, agreeably to what was advanced in a former Essay*, it must, I conceive, appear, that in proportion as man's *moral* superiority is held to consist in attributes *not* of a *material* or corporeal kind or origin, it can signify little how his *physical* nature may have originated. The same moral superiority may equally belong to him whether originally evolved out of any form of earlier organic life or out of a clod of earth. All truths relative to man's moral or spiritual nature, in proportion as that nature is held to be of an *immaterial* kind, must be allowed to be entirely independent of any theories of the origin of his animal existence.

Physical origin of man independent of his moral nature.

Objection with regard to the primeval state of man.

The difficulties felt on this subject by some seem mainly to arise out of the belief with respect to

* See Essay I. § II., and Essay II. § II.

man's primeval state of innocence. But the Scripture account, it will surely be admitted, altogether refers, *not* to man's *physical constitution*, but to the peculiar *spiritual* nature given to him; expressly described as "breathed into him"* by a special act, and which is generally conceived by divines to have constituted "the image of God,"† in which he was made; in which he stood, and from which he fell; all which can surely in no way be affected by what may have been his animal nature or origin *prior to that spiritual creation;* as it refers to that part of his nature which is spoken of expressly as distinct from, and independent of, his physical constitution and material organisation.

But if we look to the New Testament view of the matter, it will be perceived that the Christian argument assumes man in a state of degradation and sin, from which it would elevate and transform him by the renovating power of Divine grace. As to any previous state, or the origin of that depravity, St. Paul, even in adopting the representations of the Old Testament, dwells on no details, but directs the

* Gen. ii. 7. † Gen. i. 27.

whole stress of his argument, not to the physical *history or origin* of the evil, but to enhancing the greatness of the *deliverance* from it, and points to Adam only to lead men to Christ.*

Does not really affect the doctrine.

Thus the adoption of philosophical views of orderly evolution will not be found to impugn religious doctrine. Thus the theologian can have no ground for denouncing such physical speculations as impious or subversive of scriptural truth. Were theories of development ever so well established, they would not affect those doctrines; they do not even contravene the letter of the physical representations of the Old Testament to as great an extent as all geology does, and still less do they offer any opposition at all to the more spiritualised representations of the New Testament.

Objections brought forward by Mr Hugh Miller.

To urge objections, however, on theological grounds against such theories has been a popular topic with a certain class of writers; and it constitutes the main object of a work, already referred to, which has attained a more especial reputation among those who adopt such theological views as those just glanced

* Rom. v. 20.

at, or who conceive physical theories necessary for the support of religious faith,—Mr. Hugh Miller's "Footprints of a Creator," &c. So far as the author's observations bear on real points of geology and palæontology, they are characterised by his well-known acuteness and power of illustration. Yet throughout the whole we cannot but observe that the polemical spirit and avowed theological bias with which it is written cannot but weaken the authority of many parts of his physical argument.

The author introduces with great effect remarks on the high organisation of the early fishes, on which he is so well qualified to dilate, and which he enlarges upon with so much animation as furnishing indubitable "footprints" of the Creator;—that this is so every reader will most willingly admit, but in what sense it is *more peculiarly so* than is supplied by the contemplation of any other organised structures, or why more so than if a successive development were made out, it is to my apprehension impossible to conceive.

The author, though a very strenuous theological champion, is yet candid enough to allow that the theory of transmutation, if established, would not be absolutely "atheistic;" yet he considers it hardly less destructive to religion, because he conceives (unless I mistake his meaning) it would make the human soul a part of the mere development from the material world; and again, because it is opposed to the doctrine of man's primeval innocence*, though in accordance with what has just been said, in what way it could affect the former doctrine more than the fact of man's natural birth from a parent does, or how it is at all related to the latter tenet, is equally difficult to perceive.

Origin of objections in narrow Judaical views.

There are many who in their zeal for the authority of the Old Testament overlook, or, indeed, seem altogether ignorant of, the better views disclosed in the New; and it is to this Judaical school that Mr. H. Miller's theology seems traceable, as appears in a more special manner from some remarks towards the conclusion of his volume. But, without diverging into irrelevant particulars, I would only observe, with regard to the pervading principle of this work, that the author appears to consider the entire question as one between the idea of Divine operation, and what

* Footprints, &c., p. 17.

by a very common confusion of thought he regards as antagonistic to it, the principle of natural law and order, and censures "those who would transfer the work of creation from the department of miracle to the province of natural law, and would strike down in the process of removal all the old landmarks ethical and religious,"* as if the great principle of natural laws and the order of physical causes were not as entirely the emanation of the Supreme Mind, as any supposed intervention could be, and, in fact, the only true proof of it.

In no way affecting Christianity.

It can only be in the want of a more worthy appreciation of the true nature and real independence of Christianity that we hear denunciations of more enlightened views as "removing the landmarks of all religious and moral obligations," because the authority of the Old Testament is thus invaded, which, however, a slight attention to the language of the Apostle of the Gentiles must convince any unprejudiced inquirer has been in modern times unhappily mixed up with Christianity *in a way directly opposed to the whole spirit and tenor of his teaching.*†

---

* Footprints, ib.

† The practical influence of these Judaical views, displayed in the

Nor can it fail to be observed how miserably low must be the notions entertained of the grounds of *moral* obligation, if they can be supposed implicated in a question as to the process or order of the *physical* creation. Whether those grounds be regarded as connected with intuitive and immutable natural principles, or whether they be referred to the simple authority of Christian precepts, built upon faith, they must be equally independent of all theories of *creation* and of the *Judaical law.*

Successive creations made an argument for miracles.

In connexion with this subject, one other argument may here be noticed because it has been dwelt upon by some writers of eminence — the application of the supposition of "Successive Creations," in the sense of interruptions, in support of the belief in *miracles* generally; but this argument (apart from the hypothetical nature of the events assumed) will easily be seen to be of very little force, when we recollect that these so-called "creations" were, by

spirit of Sabbatism, is sometimes unhappily exercised even over science. Humboldt justly satirises "the English Sunday, on which it is sinful after Saturday night at twelve o'clock to read off a scale," as having destroyed the value of an important set of magnetic observations. (Cosmos, 1st transl., note 113 to p. 188.) Yet extensive tables of certain observations are still printed in which every *seventh* entry, instead of degrees, minutes, and seconds, is filled up by the word "Sunday!" It would be a curious calculation to find the real value of a *mean* deduced from such a column!

supposition, events *constantly recurring, and essentially different in their entire nature and circumstances* from any alleged miracles, wholly unconnected with any *revelation*, and according to the very terms of the assumption, they were the commencement and establishment of a series of *natural* results, of which (according to the view commonly adopted) miracles are professedly the violations.

Those, indeed, who think it more satisfactory to adopt that view of miracles, which has obtained the sanction of so many eminent and orthodox divines *, — that instead of interruptions they are really to be regarded rather as instances of the observance of some more comprehensive laws unknown to us, — will of course see little value in such an argument as that just referred to, but will naturally feel it much more congenial to their ideas to fall in with the more elevated conceptions of law and order presiding over

* This view has also been remarkably elucidated by Mr. Babbage (ninth Bridgewater Treatise, ch. viii.) from the nature of "laws intermitting," as exemplified in several parts of mathematical analysis and in his own calculating engine — that is to say, a mathematical formula, or a series of mechanical movements — is originally so constituted and framed that it shall give a long series of results of one continuous character, but at some one point shall exhibit a singular apparent interruption of that series or deviation from it which is nevertheless really as much a part of the series as any of the more regular terms.

even the earliest changes and evolutions of the organic world.

It is probably in reference to the species of argument just mentioned that Mr. H. Miller expresses his opinion that in the present age "the battle of the evidences will have to be fought on the field of physical science;"* but if it be on the fair field of true inductive philosophy the victory will clearly be on the side of law, design, arrangement, and subordination of causes, as the true exponents of supreme creative power and wisdom, and the real evidences of *natural* theology.

Does not affect the evidences of Christianity.

With regard to those of Christianity, they will always mainly rest on the internal and moral proof which it carries with it; and the more the age advances in real enlightenment the more will its purely spiritual claims be evinced, as wholly independent of those adaptations of an earlier dispensation, restricted, as they were suited, to the condition, the ignorance, and wants of a particular people in long-past ages; but which, nevertheless, even at the present day, are still by many strangely regarded as

* Footprints, p. 21.

if they were designed for permanent and universal truth.

The Christian doctrines, from their very nature, are conveyed in the language of the spiritual world; they belong altogether to a higher order of things; and where they may be expressed as in any degree related to material objects or events, these representations cannot now be canvassed in detail, nor be interpreted as falling within the province of physical investigation, or as involving what are described as supernatural mysteries, in any alleged physical difficulties. The truths they embody shine calmly by their own heavenly light, like the stars above the brightest illuminations on earth.

All external evidences must necessarily vary in their nature, force, and application, as addressed to different ages and persons of different capacity; unless so adapted, they must fail in their object. And this accords with what we find was the actual method and practice of the founders of Christianity in their appeals to the different parties and classes of minds they addressed.

The evidences of natural theology (such as they were in that age) are expressly recognised by

Christianity*; and it is, therefore, in entire accordance with its spirit that we follow them out at the present day in any more extended speculations to which we are led by improved science, and by which we may be able better to elucidate the order and method of creation.

Natural theology progressive with science.

Though it is the attribute of Divine truth to be one and the same for ever, it is no disparagement to that invariableness that natural theology should be *progressively changing* in the *aspect* and character of its evidence, with the improvement and advances of those sciences on which it is founded; and thus leading us to *more enlarged and worthy conceptions* of the Infinite and Supreme Intelligence, which cannot but exert a beneficial influence on the views we subsequently form of more particular doctrines. Thus to shrink from any investigation because it may seem to disparage hitherto accepted ideas, or to unsettle old convictions, is a mere mark of weakness and timidity which is inconsistent with the resolute pursuit of truth, and can end in nothing but endangering the very cause we seek to serve,

* See Rom. i. 20.; Acts xiv. 17.

and yielding up the vantage ground to the opponents.

To recapitulate and conclude: as in the *existing* condition of the material world, in those phenomena which are best understood and most perfectly investigated in all their laws and relations, it is that we have the highest and most indisputable evidences of the Supreme Intelligence; so in regard to the *past* in the same way, where we can best trace the steps and processes by which the changes have gone on, there we recognise the true evidences of creation. Yet it is the very reverse of this view which a certain class of writers would seem to uphold. They would seek the proofs of creation, not in the *known*, but in the *unknown*, regions of Nature; and precisely in those instances where we are least able to trace order and system in the Divine design, there they think we should most properly find its evidence! that we should acknowledge its proofs rather in the ignorance than in the knowledge of those recondite laws by which its reason is manifested! that we should behold the Deity *more* clearly in the *dark* than in the *light*; —in confusion, interruption, and catastrophe, *more* than in order, continuity, and progress!

Conclusion.

Evidence of creation not in interruption but in orderly processes.

Parallel in human governments.

If in travelling in a strange country we see around us the signs of order and security, civilisation and improvement, law and justice, maintained without the violent or visible interference of authority or force, we immediately infer that we are in the territory of a firmly-established, wise, and beneficent government. But if we pass into another district, where all these signs are wanting,—where we find the country in a convulsed and tumultuous state, the social machine unhinged, and the arm of power displayed only in coercion,—if we should be told that this district was subject to the same government, we might reasonably doubt or deny it; assuredly we should not infer it. But most certainly, if the lawless and convulsed state of the country and the display of arbitrary force were to be pointed out to us as the *special proofs and indications* of such dominion;—if our guide were to say, "Behold here the proper display of the majesty of the law,—here the true evidence of the greatness and wisdom of the ruler, of which you see nothing in the peaceful territory,"—we could only regard it as a mockery. Yet such is the argument of "The Footprints," and other popular works of the same class.

Imagined interruptions of preordained order for the introduction of new forms of life, so far from evincing perfections, must appear rather like blemishes in the beauty of creation; marring the picture by blots or blanks, where we should fail to follow the outline or trace the artist's design,—however we might, from the surrounding parts, conjecture its continuity.

Interruptions of law interruptions of evidence.

If, in following the track of a person, we for a time lose sight of it in broken ground, and afterwards regain it, it would be absurd to say that we *discover his footsteps* in the broken ground, because we may *infer* that he must have passed over it; so we see the rational evidence of creative power and wisdom wherever we can trace the particular steps, laws, and processes by which its operations have proceeded; but where the order of causes may be as yet hidden from us, to say that there *especially* we recognise the "indications" and "footprints of the Creator," is a contradiction. These are exactly the points in which those indications are *wanting.*

Evidence of footprints.

Through all past time we discern *everywhere* the footsteps of the Creator,—in all extinct as well as all living organic structures, modelled upon one

Wanting only in the origination of new species.

plan,—in their marvellous affinities, all linked in one chain,—in the whole scheme of one continuous series of causes in which they are united,— everywhere, *except only* in the *mode* of their *origination*, because that is not as yet traced to its law; *there alone*—these *footsteps* are, in consequence, at present *concealed* from us, though even there analogy points to them through the principle of orderly evolution.

The beginning a matter of revelation.

Science demonstrates incessant past changes, and dimly points to yet earlier links in a more vast series of development of material existence; but the idea of a *beginning*, or of *creation*, in the sense of the original operation of the Divine volition to constitute nature and matter, is beyond the province of physical *philosophy*, and can only belong to that of *faith*, and find expression in the language of *inspiration*.

Order the Divine archetype.

But though we know not what was the beginning, or will be the end, of created things, and though the whole of the present, equally with the past, be but changing phases of existence, and the material universe itself be but perishable and transitory,—yet LAW and ORDER existed before them, and will continue after them; HARMONY and SYMMETRY are permanent and eternal,—the archetypes of the Divine plan,—the very impress of that Supreme REASON and "WISDOM,"* which "was set up from everlasting;" which "when He prepared the heavens was there," and "when He appointed the foundations of the earth was by Him;" the Divine LOGOS of the Christian Genesis†, who "was in the beginning with God, and who was God: by whom all things were made, and without Him was not anything made that was made."

* Prov. viii. 23. 29. † John i. 1.

APPENDIX.

APPENDIX.

No. I.

On Inductive Reasoning.

Note to pp. 6. 13.

The well-known passage in which Aristotle* analyses the logical nature of inductive proof, and its relation to syllogism, is confessedly not only obscure, but seems to involve a contradiction, especially in his contrasting syllogism and induction in one place, and yet showing that induction may be reduced to syllogism in another. The difficulties of the case have been elaborately discussed by Dr. Whewell.† Aristotle's view is simply reducible to this,—that when an inductive argument is put into the form of a syllogism, it necessarily involves, as the major premise, the assumption that *all* objects of the kind, of which *some* are enumerated, are *like those enumerated* in the particular respect specified. With this assumption formally introduced, the syllogism is perfect in point of form. This is exactly what is pointed out by Archbishop Whately.‡ Or, in other words, it

* Anal. Prior. ii. 25.

† Camb. Trans. ix. pl. 1.

‡ Logic, bk. iv. ch. i. § 1.

amounts to saying that the argument of induction cannot be reduced to a syllogistic form except by formally making this assumption.

The only real question involved is, as to the means of arriving at the *truth* of this assumption. But it is a main consideration, often overlooked, that the *truth* of the *premises*, or the source whence we derive them, is, in the language of logicians, a matter wholly "*extra dictionem*," and with which the syllogistic theory, as such, is no way concerned. But if we proceed to consider the origin of this assumption, it is no doubt arrived at by a process of reasoning and abstraction. Aristotle says it is necessary (νοεῖν) to conceive it by an intellectual act. The question is, as to the nature of this act; and this is what I have above endeavoured to elucidate.

Mr. Mill questions the principle of this assumption, and, in some respects, opposes the views of Archbishop Whately*; yet seems to admit that every induction is or is not valid, according as the particular instances adduced are, or are not, sufficient to make it allowable to draw the general inference. But it may be asked whether, to decide this in the affirmative, is not, in fact, equivalent to making the assumption in question?

It is also important to bear in mind another distinction (which Mr. Mill himself has elsewhere admirably illustrated and insisted on†), viz., that, between the *de facto* origin and sources of our convictions, on the one hand, and the *logical order* into which they *may be analysed* on the other, they differ as a physician's prescription differs from a chemical analysis of the medicine.

* Logic, i. 373.

† Ib. i. 267.

With respect to the peculiar "inductive principle" maintained by some, it is alleged that we have a certain internal power of conception, which nevertheless requires for its operation the exhibition of a certain amount of external facts; and this has been eloquently compared to the inherent powers of vegetable life, as in a bud, to develope to a flower; yet the external influences of sun, rain, &c., are not less necessary to its action. Or, again, in the same way, the eye, constructed as it is, could not see without light; or, otherwise constructed, could not see in the light.*

But all these illustrations, apt and imposing as they are, are, after all, of little real force, if the first assumption, that we *have* such a distinct internal power, is shown to be unnecessary; and that there is nothing really peculiar in the case; as has been above attempted.

The general assumption that the mind has a power of inferring with certainty, more than actual experience warrants, presented in a variety of forms of illustration, forms the substance of various speculations on this question; all which, I believe, are capable of analysis, and therefore ought to be subjected to it, up to simpler principles. The favourite practice is to avoid this labour, by setting down everything as a *peculiar principle.*

Thus, Descartes speaks of the "seeds of truth which exist naturally in our souls."† Others have contended for innate "capacities," not ideas, and have asserted a combined action of these capacities with experience of the senses, as leading to inductive generalisation.

* See an able article in the Edinb. Review, Jan. 1852, p. 28.; also De Morgan, Formal Logic, p. 32.

† Methodé, p. 5.

A proof of such intuition is often alleged in mathematical axioms, or in the deduction of *necessary* truth generally, on which I have commented at large in my Essay on Contingent and Necessary Truth before cited, and I conceive have indicated that they *may be* reduced to simpler elements requiring no such theoretical assumption.

No. II.

Note to p. 67.

On the question to which I have referred *, respecting the supposed peculiar vital principle, I have great satisfaction in referring to various productions of Dr. Carpenter, especially his article "Life," in the Cyclopædia of Anatomy and Physiology; his Essay on the Mutual Relations of the Vital and Physical Forces †, as well as to his Principles of General and Comparative Physiology, 3rd Edit. Chap. iii. Dr. Carpenter has not only shown that the principle of the "Correlation of Forces" may be applied to those concerned in the production of *Vital* phenomena, but has recently argued for its extension to *Mental* operations, in so far as these take place automatically, *i. e.* independently of the Will. See his Principles of Human Physiology, 5th Edit., Chap. xi. Sect. 6.

No. III.

Note to p. 93.

The anomaly of retrograde motion presented by the satellites of Uranus, has been very recently shown to

* In Essay I. § II. p. 67.
† Philosophical Transactions, 1850.

extend to the satellite of Neptune, though much less highly inclined, from the observations of Mr. Lassell, as discussed by Mr. Hind.* Thus the *anomaly* is likely to cease to be one, and to become a part of some greater law, affecting in this manner the *outer* planets of our system; and it would seem to point to some cause acting exteriorly to our system while yet in a nebulous state.

It is also conceivable that the motion of the solar system through space may bring the component bodies of it into contact with other portions of cosmical matter; as indeed was suggested by an eminent continental astronomer, as the means by which new comets are continually brought within the range of our sun's attraction, and ultimately fixed in our system.

No. IV.

Note to p. 110.

The interesting little volume of Psychological Enquiries recently published by Sir B. Brodie throws valuable light on many of the topics noticed in these Essays. I regret not to have seen it till a great part of this volume was printed.

The highly curious subject of the connexion of our physical and mental constitution which forms a main topic, and is so copiously illustrated through the whole series of these "Enquiries," will throw much light on the points hinted at above.† And the remarks on the comparative endowments of man and inferior animals,

* Astron. Society's Notices, vol. xv. p. 46.
† Essay I. p. 110.

full of the most profound interest, will elucidate many of the questions here referred to; but especially the facts mentioned* may have an important bearing on what is here hypothetically suggested† on the relation of man to the system of nature.

No. V.

Note to p. 132.

In addition to what was observed before on causation, it is somewhat curious to notice that, on the other hand, D. Stewart‡ has adduced this very doctrine of mere observed sequence (discarding the notion of *necessity*), as furnishing the most effectual reply to Spinoza's material and atheistic theory derived from the supposed NECESSARY *connexion* of causes and effects throughout nature.

"Necessity" is evidently here spoken of in the confused and mystified sense once adopted of something inherently *fated* and independent of arranged order, reason, or moral causation. Hobbes' theory of religion is pervaded by the same confusion of ideas with respect to the word "cause," applied indiscriminately to physical causes and moral, above dwelt upon.

No. VI.

Abstract of Professor Owen's View of Vertebral Structure and its Archetype.

Referred to, p. 370.

The investigations of Professor Owen, especially as delivered in his essay "On Limbs," referred to in the text,

* Enquiries, pp. 172—179.
† Essay II. p. 246.
‡ Prelim. Diss. p. 210.

are so important, that it may be highly desirable to subjoin a somewhat more detailed analysis; in drawing up which, it will be no small recommendation to state, I have had the benefit of the author's own revision and remarks.*

On a cursory view, the skeleton (especially in the higher animals) appears to consist of a chain of vertebræ, terminated by the tail or sacrum at one end and the cranium at the other; while to a portion of the vertebræ are attached ribs, to the sacrum the pelvis, and to it the lower extremities, and (apparently without any connexion with vertebræ) to the upper ribs the scapula, and clavicle with the anterior extremities.

In the attempt to reduce all these parts to one principle of analogy, Cuvier, Carus, and others, made some advances. The bones of the sacrum were shown to be properly included in the class of vertebræ. But the most remarkable was the idea of reduction of the bones of the cranium, under the same analogy, at first merely as an hypothesis, by Oken.

Again, the relations of the pelvis and its limbs, and especially of the scapula and fore-extremities, were still not included in the generalisation. They had been referred to imperfect, or even positively incorrect analogies, as being liberated ribs, &c., or even expressly set down as anomalies, by Cuvier, Carus, Geoffroy, and others.† The obscurity chiefly arose from studying too exclusively the higher types, whence the nomenclature was formed on too limited a basis.‡ The comparative anatomy of lower forms suggests the true analogy.

* See also the same author's Lectures on Vertebrate Animals, 1846; on the Archetype and Homologies, &c., 1848.

† On Limbs, pp. 31. 41. 53. 102.

‡ Ib. pp. 55. 115.

Now, as to the cranium, had the idea of Oken been supported by the requisite *proofs*, the whole vertebral column would thus have been included in the same analogy; but, being hypothetical only, Oken's views were opposed by Cuvier and Agassiz, and had become virtually excluded from anatomical science at the period of the communication of Professor Owen's Report on the Homologies of the Vertebrate Skeleton to the meeting of the British Association at Southampton, in 1846. In this the generalisation was revived and established.

Again, as to the vertebræ and limbs, Professor Owen cleared up the difficulty by commencing with a more accurate view of the *nature of a vertebra* as a segment consisting essentially of a "centrum"* from which certain "apophyses"† radiate; on one side uniting to form the channel through which the nervous system of the spine is conveyed (thence called the "neural arch," or "neural apophysis"); on the other, the usually larger arch, which includes the blood system, viscera, &c., thence called the "hœmal arch," or "hœmal apophysis." These apophyses in some vertebræ take the form of *ribs*, and are here termed the "costal arch." But in different vertebræ these apophyses are differently developed; in some instances only rudimentary, or only one or two parts more developed, according to the position and organisation of the part.

But the most essential point (which could never be

* On Limbs, pp. 43. 81.

† Professor Owen restricts the term "appendage" to the part articulated to and diverging from the "apophyses," whether *neur-*, *pleur-*, or *hœm-apophyses*.

discovered but by comparative anatomy of lower forms, and by tracing the development of the higher) is that the arches are *often displaced* * from their vertebræ — sometimes to a greater, sometimes to a less extent; and that to certain arches are attached appendages which diverge from them.†

In this way the bones of the pelvis and those of the posterior extremities are shown to be the developed hœmal arches and appendages of the vertebræ of the sacrum. In the lower forms (as in fishes and serpents) the rudiments of those extremities are found, but unattached to their proper vertebra or segment. In other cases they approach in different degrees towards the condition of attachment and full development.‡

The *occipital vertebræ* are the only ones which in the higher forms appear destitute of a costal arch and appendages *in situ;* but in lower forms (as in fishes, and especially in the Lepidosiren) the arch *is seen to be formed by the scapula and clavicle,* which arch is more displaced in the crocodile, and still more in the mammalia, but the true analogy is thus seen. *The scapula, clavicle, and fore limbs are the hœmal arch and appendages of the occipital vertebræ* §, but differently displaced and developed in different orders by adaptive power.

The same differences in the development and displacement of limbs, according to this analogy, are also shown in the stages of the fœtal growth in the higher classes. ||

The undeveloped appendages of other vertebræ are

* On Limbs, pp. 50. 61.
† Ib. pp. 78. 105. 116.
‡ Ib. p. 58.
§ Ib. p. 69.
|| Ib. p. 99.

potential or rudimentary limbs, of which examples are found in fishes. *

Thus the whole skeleton is reduced into one single scheme or archetype most resembling the fish form. In different instances the parts are differently modified, but always in accordance with one invariable type or system.

As to the insufficiency of the narrow view of final causes, several striking instances are adduced.

To take a single instance, nothing can be more at variance with the doctrine that organs are constituted *merely with reference to the purpose they are to answer*, than the fact that the bones which in the human hand and arm have their extended development †, nevertheless exist in precisely the same number and arrangement, though altered in form, "buried up to the claws in a sheath of tough skin" in the "short trowel of the mole," and "hidden beneath the common undivided sheath of the fin of the dugong or whale." This "offers, perhaps, the most striking and suggestive instance of an *adherence to type, necessitated*, as it would seem, notwithstanding the absence of all those movements and appliances of the limb that explain the presence of the several segments on the principle of final causes in the horse and in man."

In like manner, the peculiar jointed arrangement of the bones of the great toe, suited to the purpose of a fulcrum, is strictly preserved in the bones of the foot of the elephant, though all enclosed in one massive hoof, and in the webbed hind-paddle of the seal.

"I think it will be obvious," the author observes,

* On Limbs, pp. 60. 65.

† Ib. pp. 13, 14.

"that the principle of final adaptation fails to satisfy all the conditions of the problem.

"A final purpose is indeed readily perceived and admitted in regard to the multiplied points of ossification in the skull of the human fœtus and their relation to safe parturition. But when we find that the same ossific centres are established, and in similar order, in the skull of the embryo kangaroo, which is born when an inch in length, and in that of the callow bird that breaks the brittle egg, we feel the truth of Bacon's comparison of final causes to the Vestal Virgins." *

No. VII.

On the Theory of Unity of Composition, by T. H. Huxley, Esq., F. R. S., &c.

Referred to, p. 374.

"In order to a strict classification of animated forms, we must observe, that living beings not only *are*, but they *become;* not only have they a definite structure in their adult condition, but each takes a definite road—passes through a definite succession of stages—in attaining that condition.

"It is therefore clear that the naturalist must not only make out the resemblance of their *adult structure*, but also the resemblance in *nature and order of the successive stages through which they pass.*

"For it is obvious that two living beings *might* have a similar structure in their adult condition, and yet have passed through different stages of development in attain-

* On Limbs, p. 39.

ing that condition; so that the naturalist who classed them together on the ground of their adult condition alone might be altogether wrong.

"To take an example:—

"An error in classification of this kind was made by Cuvier himself. The Cirripedes, or Barnacles, are creatures which in their adult condition present a certain resemblance to Mollusks in many of their structural characters. Cuvier, who knew them in this condition only, did not hesitate to classify them with the Mollusca.

"Later investigators, however, who have studied the entire development of the Barnacles from their youngest state upwards, have shown, that at first they are entirely similar to the Waterfleas and Monoculi of our ponds and ditches, which are Annulose animals; and that all the features in which they resemble Mollusks arise from subsequent modifications of their mere external form.

"There is no doubt, therefore, in the mind of any anatomist of the present day, that the Cirripedes are *Annulose* and not *Molluscous* animals. The error of the anatomical method has been corrected by the application of the developmental method.

"In the main, however, and perhaps invariably, when sufficient care has been exercised, the anatomical and developmental methods furnish perfectly harmonious results. Animals possessing similar adult structure, as a rule, pass through similar stages of development, and therefore it has been found that those grand generalisations of purely anatomical facts—upon which Cuvier founded his quadripartite division of the animal kingdom—have been, eventually, only confirmed and placed upon an irrefragable basis by inquiries into development.

"It cannot be too forcibly borne in mind in estimating the value of Cuvier's system, that he aimed not at a morphology, but at a classification; he did not attempt to discover upon what plans animals are constructed, but to ascertain in what manner the facts of animal organisation could be thrown into the fewest possible general propositions. He set himself to find out what structural resemblances were the marks of the greatest possible number of other structural resemblances; and having found that the similarity of the structure of the nervous system was a mark of more resemblance of other kinds than any other sort of similarity, he adopted it as the base of his great divisions.

"By such a method, without the study of development, you may have a *classification* of animals, but no *morphology;* you have *without development* no criterion of the truth or falsehood of any doctrine regarding a common plan or archetype.

"What Von Bär did was to generalise the facts of development in precisely the same way as Cuvier had generalised the facts of structure, and to demonstrate that the classification of Cuvier was in the main simply the expression of the fact that there are certain Common *Plans of Development* in the animal kingdom;—that there is one Common Plan followed by all Vertebrate animals, another by all Mollusca, a third by all Annulosa, and a fourth by all Radiata.

"Finally, the grandest law of all at which Von Bar arrived was, that although beyond *a certain period* in its existence every Vertebrate, Mollusk, Annulose, and Radiate animal followed its own special plan, yet that *up to that point it followed a plan common to all animals;*

and thus he demonstrated and placed upon a footing as secure as that of the law of gravitation that doctrine of the unity of organisation of all animals, which was with Geoffroy an undemonstrated hypothesis.

" In a word, the leading idea on which the doctrine of a 'common plan' now rests is the possibility of demonstrating a *common mode of development* for those animals which are affirmed to be organised upon a Common Plan :—*Development is the Criterion of Homology.*"

No. VIII.

Abstract of the Theory of Specific Centres. By the late Professor E. Forbes, F. R. S., &c.

Referred to, p. 403.

" 1. A 'specific centre' is an area occupied by the individuals of a species.

" 2. It is an ascertained fact, that numerous well-marked *provinces* of the earth and sea can be indicated, each characterised by a flora and fauna on the whole peculiar to itself.

" 3. A *species* absolutely peculiar to a *province* has necessarily its centre within it; but many species are common to two or more provinces.

" 4. As a rule, when a species is common to two or more provinces, these provinces are contiguous, and, consequently, the specific area is continuous.

" 5. But there are exceptions, such as species, or a group of species, exhibiting in some cases the phenomenon of occupying more than one area, or of presenting outliers of individuals separated from the main assemblage.

" 6. But when we sift the history of such exceptions we find, that by tracing back the history of the distribution of the species, or group of species, so situated, in time (*i. e.* their geological history), we can show the strong probability of an epoch when all the individuals of the species in question occupied a continuous and unique area.

" 7. Hence an inquiry into the distribution of the individuals of a species in both time and space results in the maintenance of the theory of the *unity of specific centres.*

" 8. Moreover, when we are able to trace the history of a species in time, we find, in the majority of instances, that there is a distinct indication of a paucity of individuals as we approach the epoch of its first appearance.

" 9. And when we are dealing with well-marked and continuous areas of species belonging to the present epoch, we find that there is within such area the indication of a point of maximum development of individuals, around which the numbers diminish.

" 10. We infer from these facts (8. and 9.) the probability of a single point of origin for every species within its centre of occupation.

" 11. In the course of time, however, it is possible that the area of occupation of a species may become removed from the point of origin, or may, after being removed, eventually return to its original position.

" 12. The indications of a single point of origin for each species, combined with the fact that we have no knowledge or experience of the individuals of any species being produced otherwise than from individuals of its own kind—in other words, that we have no knowledge of any other relationship between the individuals of a species than that of *descent*—leads to the inference that each species originated from a unique stock or prototype, consisting of a single being, or pair of beings, according as would be required for propagation.

"13. Hence the point of origin within a specific centre is the point of appearance of the prototype.

"14. How that prototype originated we know not; but the doctrine of specific centres, originating each with a prototype of its own, is necessarily opposed to the hypothesis of *the evolution* of all species from one first form, without respect to the superiority or inferiority of the form.

"15. That the prototype presented the *specific characters* (*i. e.* distinctive and constant features) of its descendants, is an hypothesis rendered probable by all that we know of the history of species in space and time.

"16. The observation of the distribution of species in space and time indicates geographical areas and chronological epochs—points in time and space—where, as it were in preference, many species originated in groups. These we term *centres of creation*, and the phenomena of provinces are linked with the existence of them.

"17. The value of Palæontology to Geology depends on the assumption of the constancy of specific types, and the unity of their centres or areas of occupation in time.

"18. What is true with existing species should be *à priori* true with extinct ones, since we can clearly show that all known creatures, recent and fossil, are members of one biological system."

No. IX.

On the recent Origin of Man.

Note to p. 464.

A discovery of fossil *human remains* has been recently made under circumstances which appear to me to call for much more close examination than (as far as I am aware) appears to have been bestowed on the case.

The statement to which I refer is given in a paper on "The Railway Cuttings at Mickleton Tunnel," &c., by G. E. Gavey, Esq., Quarterly Geological Journal, February, 1853, p. 32., where the fact of the occurrence of these remains is passed over without any comment.

The annexed sketch is taken from that accompanying the paper, the *proportions* only being exaggerated for clearness.

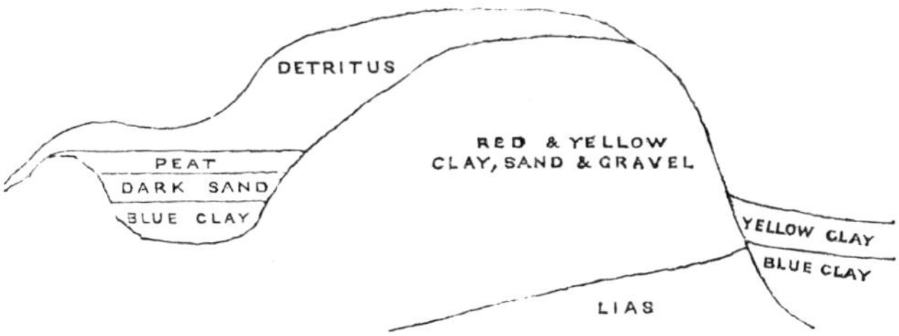

The material facts are briefly these:—The section of the hill presents a basis of lias, immediately over which is a mass of *yellow and red* sand, clay, and gravel, forming the summit; on the northern side, after a mass of superficial detritus, there was displayed what had been a small hollow or depression in the slope of the hill, the upper part of which was filled up by an ancient peat bog, containing bones of various animals, probably of existing species, deer, oxen, &c.; below this occurred a bed of dark sandy earth, with stumps of trees, *in situ;* and below this a bed of indurated blue clay (the same as that forming a continuous band on the opposite face of the hill), in which a *human skeleton* was found, in an inclined position, distinctly as *engulphed, not buried.* Now on this I would observe:—

1. The beds composing the upper part of the hill consist solely of *yellow* and *red* clays, and sands, &c.; hence the *blue* clay *could not* have come from the upper part of the hill by washing down at a recent period, but could only have originated in a deposition from the waters before the hill had emerged to the level at which it occurs, and which is nearly the same on both sides.

[The *yellow* clay which covers it on the south *may possibly* have been washed down, but may more probably have been a similar deposit.]

2. After the hill had emerged, the *blue clay* on the south was dried up and consolidated. But on the north the small patch of it continued exposed, and did not dry up or consolidate, but remained in the state of *soft mud* probably by the retention of water in the hollow, at least until the *human remains* were inclosed in it, which must have taken place while it was in a soft state (unless, indeed, they were engulphed while it was still under the sea, which is not likely). *How long after the elevation* it thus continued in the state of a muddy pond before being filled up and coated over, is another point not easily settled.

3. At some period *after* these remains were imbedded, the water drained off, the clay consolidated, and a bed of dark sandy earth overspread the top of it. Whence was this derived?

4. In this, *trees* then grew, their roots striking down into the clay; their stumps being found *in situ*, with various other vegetable remains in the sand.

5. After the trees had fallen and decayed, a peat bog formed on their site. All these events must have required a long series of ages. The peat bog, when at length formed fully, must have remained such for a very long period to account for the great *number* of remains of various animals — oxen, horses, deer, boars, and foxes — all collected in so small a space.

6. After this the peat dried up, and the whole became covered with a deposit of loam, sand, &c., which overspreads the side and top of the hill, but whose date is not explained, and which forms the modern surface and contour of the hill.

Considering the very long series of physical events which thus *must* have occurred *since the human remains were imbedded*, it becomes an important inquiry to endeavour to settle the probable relation of these various changes to any known epochs of geological action.

I merely wish to place these facts in a prominent light for the sake of exciting inquiry on the part of those better able to judge, and without pretending to offer an opinion on the point. The question would probably involve attention to the repeated series of changes of level and condition which have occurred in the long period since the pleiocene, especially as illustrated by the researches of Mr. Trimmer.*

* See Geol. Quart. Journal, No. 36. p. 293., and previous numbers there referred to.

THE END.